Water Resources and Environmental Depth Practice Exams
for the Civil PE Exam

R. Wane Schneiter, PhD, PE

PPI®
PPI2PASS.COM

Professional Publications, Inc. • Belmont, California

Benefit by Registering This Book with PPI

- Get book updates and corrections.
- Hear the latest exam news.
- Obtain exclusive exam tips and strategies.
- Receive special discounts.

Register your book at **ppi2pass.com/register**.

Report Errors and View Corrections for This Book

PPI is grateful to every reader who notifies us of a possible error. Your feedback allows us to improve the quality and accuracy of our products. You can report errata and view corrections at **ppi2pass.com/errata**.

WATER RESOURCES AND ENVIRONMENTAL DEPTH PRACTICE EXAMS
First Edition

Current release of this edition: 3

Release History

date	edition number	revision number	update
Oct 2014	1	1	New book.
Jun 2015	1	2	Minor cover updates.
Nov 2016	1	3	Minor corrections. Minor formatting and pagination changes. Minor cover updates.

© 2014 Professional Publications, Inc. All rights reserved.

All content is copyrighted by Professional Publications, Inc. (PPI). No part, either text or image, may be used for any purpose other than personal use. Reproduction, modification, storage in a retrieval system or retransmission, in any form or by any means, electronic, mechanical, or otherwise, for reasons other than personal use, without prior written permission from the publisher is strictly prohibited. For written permission, contact PPI at permissions@ppi2pass.com.

Printed in the United States of America.

PPI
1250 Fifth Avenue, Belmont, CA 94002
(650) 593-9119
ppi2pass.com

ISBN: 978–1–59126–396–8

Library of Congress Control Number: 2012948728

F E D C B A

Table of Contents

Preface and Acknowledgments

For many, the title of professional engineer (PE) represents the fulfillment of a journey that begins in college. But, math and basic engineering courses do not make a PE, nor do upper-division courses in specialty areas. It is only when a person satisfies the criteria to use the title of professional engineer that the journey is complete. Since I passed the Principles and Practice of Engineering exam in 1984, I have earned other titles and filled a variety of engineering roles, but I proudly define myself as a professional engineer.

I have worked as a consulting environmental engineer since I became a PE. However, over most of the last two decades, I have also worked in academia. As a professor, I have tried to represent myself as an engineer who teaches rather than a teacher who is an engineer. This attitude has allowed me to present the practical perspective of engineering to my students, emphasizing the importance of professional registration, so that they too will be able to claim the title of professional engineer. From my teaching experiences, I learned what engineers studying for the water resources and environmental depth module of the civil PE exam needed—an opportunity to solve exam-like problems. I wrote *Water Resources and Environmental Depth Practice Exams* to fulfill this need. By using this book to simulate the exam, you will build the confidence and problem-solving skills necessary to pass the exam.

As engineers, you and I know that any successful project is the result of team effort. The project that this book represents was successful because of the staff in PPI's product development and implementation department. In particular, I would like to acknowledge the contributions of Tyler Hayes, copy editor, who provided invaluable editorial assistance, and of Sarah Hubbard, director of product development and implementation, for her continued interest in my contributions to the PPI catalog. There are many others at PPI whom I haven't met, but I know they have worked hard on this project. I express my gratitude to them: Cathy Schrott, production services manager; Julia White, editorial project manager; Christine Eng, product development manager; David Chu, Lisa Devoto Farrell, Magnolia Molcan, Ellen Nordman, Bonnie Thomas, Heather Turbeville, and Ian A. Walker, copy editors; Aruna Raghuram, EIT, technical reviewer; Jeanette Baker, EIT, and Scott Miller, EIT, calculation checkers; and Tom Bergstrom and Kate Hayes, production associates.

I would also like to thank the following individuals for contributing their time and engineering expertise by technically reviewing this book: David W. Johnstone, PhD, PE; Gerald J. Kauffman, PhD, PE; Aparna Phadnis, PE; and David L. Sheridan, PhD, PE.

One thing about engineering is that our work is available for our peers and the public to see. If we make errors, someone will find them. Although I have taken great care to ensure that the problems in this book are presented clearly and solved correctly, and that all assumptions are reasonable, there will inevitably be errors that I missed. Any errors, omissions, inconsistencies, or other flaws are mine, and I apologize for these. I ask that you help improve this book by sharing your corrections or comments with me by submitting them using PPI's errata website at **ppi2pass.com/errata**.

R. Wane Schneiter, PhD, PE

Recommended References

The water resources and environmental depth module of the civil PE exam is not based on specific codes or references. However, the minimum recommended library for the exam includes the *Civil Engineering Reference Manual* and the *Water Resources and Environmental Depth Reference Manual*. References that may be useful during the exam are given as follows.

Aisenbrey, A. J., Jr., et al. *Design of Small Canal Structures*. Denver, CO: U.S. Department of the Interior, Bureau of Reclamation.

Chow, V. T. *Open-Channel Hydraulics*. New York, NY: McGraw-Hill.

Fetter, C. W. *Applied Hydrogeology*. Upper Saddle River, NJ: Prentice Hall.

Fetter, C. W. *Contaminant Hydrogeology*. Upper Saddle River, NJ: Prentice Hall.

Haynes, W. M. *CRC Handbook of Chemistry and Physics*. Boca Raton, FL: CRC Press.

Linsley, R. K., et al. *Hydrology for Engineers*. New York, NY: McGraw-Hill.

Linsley, R. K., et al. *Water Resources Engineering*. New York, NY: McGraw-Hill.

Luthin, J. N. *Drainage Engineering*. Huntington, NY: R. E. Krieger Publishing Company.

Masters, G. M., et al. *Introduction to Environmental Engineering and Science*. Upper Saddle River, NJ: Prentice Hall.

McGhee, T. J., et al. *Water Supply and Sewerage*. New York, NY: McGraw-Hill.

Munson, B. R., et al. *Fundamentals of Fluid Mechanics*. Hoboken, NJ: John Wiley & Sons.

Natural Resources Conservation Service. *Urban Hydrology for Small Watersheds* (TR-55). Washington, DC: U.S. Department of Agriculture.[1]

Peavy, H. S., et al. *Environmental Engineering*. New York, NY: McGraw-Hill.

Precipitation-Frequency Atlas of the United States. NOAA Atlas 14, Vol. 1 and Vol. 1 Addendum. National Oceanic and Atmospheric Administration (NOAA).[1]

Precipitation-Frequency Atlas of the Western United States. NOAA Atlas 2. National Oceanic and Atmospheric Administration (NOAA).[1]

Recommended Standards for Wastewater Facilities (Ten States Standards). Great Lakes—Upper Mississippi River Board.[1]

Recommended Standards for Water Works (Ten States Standards). Great Lakes—Upper Mississippi River Board.[1]

Sawyer, C. N., et al. *Chemistry for Environmental Engineering and Science*. New York, NY: McGraw-Hill.

Seaber, P. R., et al. *Hydrologic Unit Maps*. Water-Supply Paper 2294. Denver, CO: U.S. Geological Survey.

Sincero, A. P., et al. *Environmental Engineering: A Design Approach*. Upper Saddle River, NJ: Prentice Hall.

Speight, J. G. *Lange's Handbook of Chemistry*. New York, NY: McGraw-Hill.

Tchobanoglous, G., et al. *Wastewater Engineering: Treatment and Reuse*. New York, NY: McGraw-Hill.

Viessman, W., Jr., et al. *Introduction to Hydrology*. Upper Saddle River, NJ: Prentice Hall.

Viessman, W., Jr., et al. *Water Supply and Pollution Control*. Upper Saddle River, NJ: Prentice Hall.

Williams, T. A. *Hydrologic Engineering Methods for Water Resources Development, Volume 6: Water Surface Profiles*. Davis, CA: U.S. Army Corps of Engineers.

---

[1]A link to a downloadable version is provided at **ppi2pass.com/CEwebrefs**.

Introduction

ABOUT THIS BOOK

Water Resources and Environmental Depth Practice Exams includes two exams designed to match the format and specifications of the water resources and environmental depth module of the civil PE exam. Like the actual exam, the exams in this book contain 40 multiple-choice problems, and each problem takes an average of six minutes to solve. Most of the problems are quantitative, requiring calculations to arrive at the correct option. A few are nonquantitative.

Each of the questions will have four answer options, labeled "A," "B," "C," and "D." If the answer options are numerical, they will be displayed in increasing value. One of the answer options is correct (or, will be "most nearly correct"). The remaining answer options are incorrect and may consist of one or more "logical distractors," the term used by NCEES to designate incorrect options that look correct. Incorrect options represent answers found by making common mistakes. These may be simple mathematical errors, such as failing to square a term in an equation, or more serious errors, such as using the wrong equation.

The solutions in this book are presented step-by-step to help you follow the logical development of the solving approach and to provide examples of how you may want to solve similar problems on the exam. Nomenclature is defined in each solution to help you quickly identify the variables used and determine appropriate units.

Some solutions include author commentary that uses the following icons for quick identification.

- ⏱ *Timesaver:* a technique or approach to reduce problem-solving time
- ✸ *Pitfall:* a common pitfall or distractor

Solutions presented for each problem may represent only one of several methods for obtaining the correct answer. Alternative problem-solving methods may also produce correct answers.

ABOUT THE EXAM

The Principles and Practice of Engineering (PE) exam is administered by the National Council of Examiners for Engineering and Surveying (NCEES). The civil PE exam is an eight-hour exam divided into a four-hour morning breadth exam and a four-hour afternoon depth exam. The morning breadth exam consists of 40 multiple-choice problems covering eight areas of general civil engineering knowledge: project planning; means and methods; soil mechanics; structural mechanics; hydraulics and hydrology; geometrics; materials; and site development. As the "breadth" designation implies, morning exam problems are general in nature and wide-ranging in scope. All examinees take the same breadth exam.

For the afternoon depth exam, you must select a depth module from one of the five subdisciplines: construction, geotechnical, structural, transportation, or water resources and environmental. The problems on the afternoon depth exam require more specialized knowledge than those on the morning breadth exam. Topics and the distribution of problems on the water resources and environmental depth module are as follows.

- **Analysis and Design (4 questions)**

  Mass balance; hydraulic loading; solids loading; hydraulic flow measurement

- **Hydraulics—Closed Conduit (5 questions)**

  Energy and/or continuity equation; pressure conduit; pump application and analysis; wet wells; lift stations; cavitation; pipe network analysis

- **Hydraulics—Open Channel (5 questions)**

  Open-channel flow; hydraulic energy dissipation; stormwater collection and drainage; subcritical and supercritical flow

- **Hydrology (7 questions)**

  Storm characterization; runoff analysis; hydrograph development and applications (including synthetic hydrographs); rainfall intensity, duration, and frequency; time of concentration; rainfall and stream gauging stations; depletions; stormwater management

- **Groundwater and Wells (3 questions)**

  Aquifers; groundwater flow; well analysis (steady-state only)

- **Wastewater Collection and Treatment (6 questions)**

  Wastewater collection systems; wastewater treatment processes; wastewater flow rates; preliminary treatment; primary treatment; secondary treatment;

nitrification/denitrification; phosphorus removal; solids treatment, handling, and disposal; digestion; disinfection; advanced treatment

- **Water Quality (3 questions)**

   Stream degradation; oxygen dynamics; total maximum daily load; biological contaminants; chemical contaminants (including bioaccumulation)

- **Drinking Water Distribution and Treatment (6 questions)**

   Drinking water distribution systems; drinking water treatment processes; demands; storage; sedimentation; taste and odor control; rapid mixing; flocculation; filtration; disinfection (including disinfection byproducts); hardness and softening

- **Engineering Economic Analysis (1 question)**

   Economic analysis

All problems on the breadth and depth exams are multiple choice. The problem statement includes all information required to solve the problem, followed by four options. Only one of the four options is correct. Each problem is independent, so incorrectly calculating the answer to one problem will not impact subsequent problems.

For further information and tips on how to prepare for the water resources and environmental depth module of the civil PE exam, consult the *Civil Engineering Reference Manual*, the *Water Resources and Environmental Depth Reference Manual*, or PPI's civil PE exam FAQs at **ppi2pass.com/cefaq**.

HOW TO USE THIS BOOK

Prior to taking these practice exams, locate and organize relevant resources and materials as if you are taking the actual exam. Refer to the Recommended References section for guidance on materials. Also, visit **ppi2pass.com/stateboards** for a link to your state's board of engineering, and check for any state-specific restrictions on materials you are allowed to bring to the exam. You should also check NCEES' calculator policy at **ppi2pass.com/calculators** to ensure your calculator can be used on the exam.

The two exams in this book allow you to structure your own exam preparation in the way that is best for you. For example, you might choose to take one exam as a pretest to assess your knowledge and determine the areas in which you need more review, and then take the second after you have completed additional study. Alternatively, you might choose to use one exam as a guide for how to solve different types of problems, reading each problem and solution in kind, and then use the second exam to evaluate what you learned.

Whatever your preferred exam preparation method, these exams will be most useful if you restrict yourself to exam-like conditions when solving the problems. When you are ready to begin an exam, set a timer for four hours. Use only the calculator and references you have gathered for use on the exam. Use the space provided near each problem for your calculations, and mark your answer on the answer sheet.

When you finish taking an exam, check your answers against the answer key to assess your performance. Review the solutions to any problems you answered incorrectly or were unable to answer. Read the author commentaries for tips, and compare your problem-solving approaches against those given in the solutions.

Practice Exam 1 Instructions

In accordance with the rules established by your state, you may use textbooks, handbooks, bound reference materials, and any approved battery- or solar-powered, silent calculator to work this examination. However, no blank papers, writing tablets, unbound scratch paper, or loose notes are permitted. Sufficient room for scratch work is provided in the Examination Booklet.

You are not permitted to share or exchange materials with other examinees. However, the books and other resources used in this afternoon session do not have to be the same as those used in the morning session.

You will have four hours in which to work this session of the examination. Your score will be determined by the number of questions that you answer correctly. There is a total of 40 questions. All 40 questions must be worked correctly in order to receive full credit on the exam. There are no optional questions. Each question is worth 1 point. The maximum possible score for this section of the examination is 40 points.

Partial credit is not available. No credit will be given for methodology, assumptions, or work written in your Examination Booklet.

Record all of your answers on the Answer Sheet. No credit will be given for answers marked in the Examination Booklet. Mark your answers with the official examination pencil provided to you. Marks must be dark and must completely fill the bubbles. Record only one answer per question. If you mark more than one answer, you will not receive credit for the question. If you change an answer, be sure the old bubble is erased completely; incomplete erasures may be misinterpreted as answers.

If you finish early, check your work and make sure that you have followed all instructions. After checking your answers, you may turn in your Examination Booklet and Answer Sheet and leave the examination room. Once you leave, you will not be permitted to return to work or change your answers.

When permission has been given by your proctor, break the seal on the Examination Booklet. Check that all pages are present and legible. If any part of your Examination Booklet is missing, your proctor will issue you a new Booklet.

WAIT FOR PERMISSION TO BEGIN

Name: _____

       Last   First   Middle Initial

Examinee number: _____

Examination Booklet number: _____

Principles and Practice of Engineering Examination

Afternoon Session
Practice Exam 1

Practice Exam 1 Answer Exam

Name: _____
     Last   First   Middle Initial

Date: _____

1. Ⓐ Ⓑ Ⓒ Ⓓ
2. Ⓐ Ⓑ Ⓒ Ⓓ
3. Ⓐ Ⓑ Ⓒ Ⓓ
4. Ⓐ Ⓑ Ⓒ Ⓓ
5. Ⓐ Ⓑ Ⓒ Ⓓ
6. Ⓐ Ⓑ Ⓒ Ⓓ
7. Ⓐ Ⓑ Ⓒ Ⓓ
8. Ⓐ Ⓑ Ⓒ Ⓓ
9. Ⓐ Ⓑ Ⓒ Ⓓ
10. Ⓐ Ⓑ Ⓒ Ⓓ

11. Ⓐ Ⓑ Ⓒ Ⓓ
12. Ⓐ Ⓑ Ⓒ Ⓓ
13. Ⓐ Ⓑ Ⓒ Ⓓ
14. Ⓐ Ⓑ Ⓒ Ⓓ
15. Ⓐ Ⓑ Ⓒ Ⓓ
16. Ⓐ Ⓑ Ⓒ Ⓓ
17. Ⓐ Ⓑ Ⓒ Ⓓ
18. Ⓐ Ⓑ Ⓒ Ⓓ
19. Ⓐ Ⓑ Ⓒ Ⓓ
20. Ⓐ Ⓑ Ⓒ Ⓓ

21. Ⓐ Ⓑ Ⓒ Ⓓ
22. Ⓐ Ⓑ Ⓒ Ⓓ
23. Ⓐ Ⓑ Ⓒ Ⓓ
24. Ⓐ Ⓑ Ⓒ Ⓓ
25. Ⓐ Ⓑ Ⓒ Ⓓ
26. Ⓐ Ⓑ Ⓒ Ⓓ
27. Ⓐ Ⓑ Ⓒ Ⓓ
28. Ⓐ Ⓑ Ⓒ Ⓓ
29. Ⓐ Ⓑ Ⓒ Ⓓ
30. Ⓐ Ⓑ Ⓒ Ⓓ

31. Ⓐ Ⓑ Ⓒ Ⓓ
32. Ⓐ Ⓑ Ⓒ Ⓓ
33. Ⓐ Ⓑ Ⓒ Ⓓ
34. Ⓐ Ⓑ Ⓒ Ⓓ
35. Ⓐ Ⓑ Ⓒ Ⓓ
36. Ⓐ Ⓑ Ⓒ Ⓓ
37. Ⓐ Ⓑ Ⓒ Ⓓ
38. Ⓐ Ⓑ Ⓒ Ⓓ
39. Ⓐ Ⓑ Ⓒ Ⓓ
40. Ⓐ Ⓑ Ⓒ Ⓓ

Practice Exam 1

1. For the reservoir piping system shown, the inside diameter of pipe A is 15 cm, the inside diameter of pipe B is 10 cm, and the inside diameter of pipe C is 8 cm. Water exits pipe C as a free jet. All elevations have the same reference point. Neglecting pipe friction, what is most nearly the pressure in pipe B along the centerline at point P?

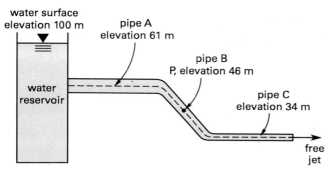

(A) 88 kN/m²

(B) 120 kN/m²

(C) 260 kN/m²

(D) 1700 kN/m²

2. A run of nominal 3 in pipe has one long radius 90° elbow at each end and is fixed in an overhead rack. The elbow diameter is also 3 in, and the diameter is constant throughout the elbow. The inside pipe diameter is equal to the nominal diameter. The elbows are in the horizontal plane, with no change in inlet and outlet elevations. Water flow in the pipe is 0.55 ft³/sec at 50 psig. Assuming frictionless flow and constant pressure throughout the pipe, the resultant force exerted by the water on the pipe rack at each elbow is most nearly

(A) 516 lbf

(B) 523 lbf

(C) 649 lbf

(D) 727 lbf

3. Two pumps are operated in series. The flow rates, total head losses, and characteristic curves for the pumps are given. What is most nearly the total discharge at the operating point of the combined pump curve?

flow rate (gpm)	total head loss (ft)
49	94.7
56	97.9
63	100.8
67	102.7
72	104.9

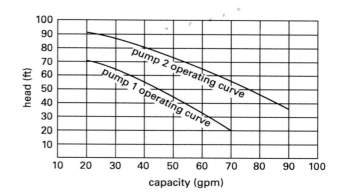

(A) 50 gpm

(B) 60 gpm

(C) 65 gpm

(D) 120 gpm

4. 70°F water flows through a 12 in diameter schedule-40 welded steel pipe that has a relative roughness of 0.0002 and an equivalent length of 1 mi. The water flow in the pipe varies between the two extremes of 2.2 ft³/sec and 4.4 ft³/sec. The loss coefficients for all flows are equivalent. If head loss is given by $h_f = kQ^n$, what is most nearly the value of k?

(A) 0.18

(B) 2.5

(C) 19

(D) 25

5. Manometers are installed near the entrance and exit of a straight pipe 20 ft long with an inside diameter of 0.75 in. The flow rate through the pipe is 8 gpm. The upstream manometer reading is 37.1 in, and the downstream manometer reading is 9.2 in. All significant head loss in the pipe is from friction. What is most nearly the specific roughness of the pipe?

(A) 0.000014 ft

(B) 0.00022 ft

(C) 0.014 ft

(D) 0.019 ft

6. Smooth schedule-40 PVC pipe is used to deliver a water flow of 20 L/s at a water temperature of 20°C over a pipe run of 320 m. The allowable head loss is 1.0 m, and minor losses are negligible. Assume an initial Darcy friction factor of 0.015 and a Reynolds number of 1×10^5. The smallest standard pipe size that will accommodate the flow is

(A) 4 in

(B) 6 in

(C) 8 in

(D) 10 in

7. Water flows in a concrete-lined open rectangular channel that is 5 ft wide. The normal water depth is 8 in. The channel has a constant slope of 0.2%. What is most nearly the flow rate of the water in the channel?

(A) 8.7 ft³/sec

(B) 11 ft³/sec

(C) 130 ft³/sec

(D) 820 ft³/sec

8. A concrete-lined open channel is used to convey stormwater runoff along a roadway. The roadway and channel make an abrupt transition from 12% slope to 2% slope, which causes a hydraulic jump to occur. The channel is a triangular cross section and has 1:1 side slopes. The upstream water depth is 10 cm. What is most nearly the water depth in the downstream channel section with the 2% slope?

(A) 0.20 m

(B) 0.28 m

(C) 0.46 m

(D) 1.2 m

9. Water flowing 3 ft deep in a trapezoidal channel enters a box culvert through an abrupt transition. The culvert extends for 1000 ft and then makes an abrupt transition back to a trapezoidal channel. The channel and culvert are constructed of reinforced concrete. The channel base width is 8 ft with 1:1 side slopes, and the culvert width is 8 ft. The channel and culvert slope are constant at 2%. What is most nearly the depth of flow in the culvert?

(A) 1.3 ft

(B) 3.0 ft

(C) 4.1 ft

(D) 5.8 ft

10. A baffled outlet is used for energy dissipation at a drainage discharge from a square channel into the basin shown. The flow from the drainage channel is 120 ft³/sec with a head of 32 ft. What is most nearly the minimum required depth, D_b, of the baffled outlet basin if its width is fixed at 6 ft?

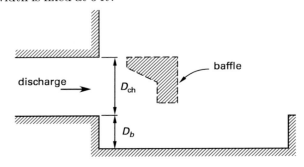

cross section

(A) 0.44 ft

(B) 0.68 ft

(C) 0.85 ft

(D) 0.91 ft

11. The height of a submerged barrier placed across the entire width of a 15 ft wide rectangular channel can be increased or decreased according to varying flow conditions. The flow velocity is 3 ft/sec, and the normal depth is 8 ft. Approximately how high can the barrier be raised before creating a change in the water depth a few feet upstream of the barrier?

(A) 1.5 ft

(B) 2.6 ft

(C) 3.0 ft

(D) 4.2 ft

12. A sluice gate is placed across a rectangular channel as an energy dissipation device as shown. The water velocity in the channel just downstream of the sluice gate is 8 m/s, and the water depth is 0.65 m. Approximately how much energy is dissipated by the hydraulic jump that forms downstream of the sluice gate?

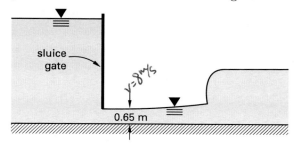

(A)   38 kW/m

(B)   57 kW/m

(C)   210 kW/m

(D)   230 kW/m

13. Intensity-duration-frequency curves for a watershed are shown. What is most nearly the frequency of a 60 min duration and 5.8 in/hr intensity storm?

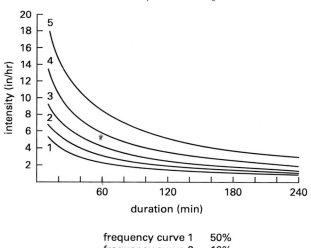

frequency curve 1	50%
frequency curve 2	10%
frequency curve 3	4%
frequency curve 4	2%
frequency curve 5	1%

(A)   2 yr

(B)   10 yr

(C)   30 yr

(D)   50 yr

14. A manufacturing facility has a 1% risk of flooding during its 50 yr design life. What is most nearly the annual probability that flooding will occur during the facility design life?

(A)   0.00020%

(B)   0.020%

(C)   0.50%

(D)   1.0%

15. The hydrograph for a storm is shown. The basin drainage area is 5480 mi^2. What is the approximate direct runoff from the storm?

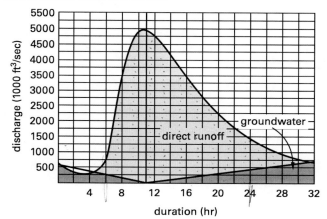

(A)   0.016 in

(B)   2.6 in

(C)   13 in

(D)   16 in

16. A slope-stage-discharge rating curve and observed fall correction curve for a stream gage station are shown. What is most nearly the stream flow for a 2.3 m stage with an observed fall of 0.3 m?

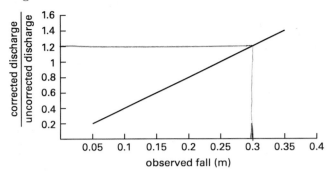

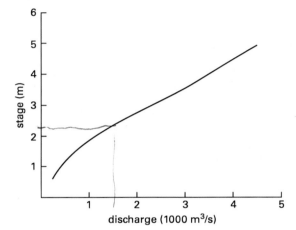

(A) 1.8 m^3/s

(B) 300 m^3/s

(C) 1200 m^3/s

(D) 1800 m^3/s

17. The area distribution by elevation in a mountainous region is shown. The snow line is at 1370 m, and the elevation of the weather station where temperature is measured is 2036 m. The ambient lapse rate is $-0.016°C/m$, and the snow melts at a rate of 3 mm/degree-day. What is most nearly the snowmelt volume for a day when the average temperature at the weather station is 12°C?

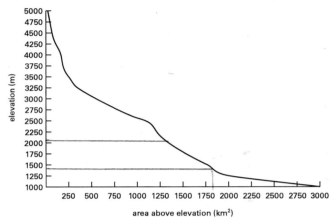

(A) 2.0×10^7 m^3

(B) 3.7×10^7 m^3

(C) 6.3×10^7 m^3

(D) 4.2×10^{10} m^3

18. A watershed consists of three adjacent tracts, each discharging to a centrally located common outlet. The area and rational runoff coefficient of each tract are given. The peak flow coincides with the time to concentration, which is 16 min. The rainfall intensity at the time to concentration is 2.8 in/hr. There is no cross runoff. What is the approximate peak runoff from the watershed at the time to concentration?

tract	area (ac)	rational runoff coefficient
1	12	0.2
2	8	0.6
3	17	0.08

(A) 2.5 ft^3/sec

(B) 24 ft^3/sec

(C) 30 ft^3/sec

(D) 91 ft^3/sec

19. An unconfined aquifer with a 4 in diameter pumped well and an observation well is shown. The aquifer is 18 ft thick and has a hydraulic conductivity of 7.2 ft/day.

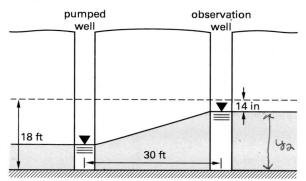

The observation well is located 30 ft from the pumped well. After 10 hr of pumping at 20 gpm, the observation well shows a steady-state drawdown of 14 in. What is most nearly the radius of influence of the pumped well?

(A) 2.3 ft

(B) 31 ft

(C) 38 ft

(D) 53 ft

20. Bore-hole logs show interbedded soil layers with the following characteristics. What is the approximate overall hydraulic conductivity for flow perpendicular to the soil layers?

layer	thickness (cm)	soil class
1	70	SP sand
2	109	GC
3	88	SM silty sand
4	46	SC clayey sand

(A) 5.0×10^{-8} cm/s

(B) 1.3×10^{-7} cm/s

(C) 1.3×10^{-4} cm/s

(D) 5.0×10^{-4} cm/s

21. In a soil-groundwater system, hydrodynamic dispersion is a combination of mechanical dispersion and diffusion. In a low permeability soil with a very shallow groundwater gradient, which component of hydrodynamic dispersion will dominate, and what will be the general shape of the solute plume?

(A) Dispersion dominates. The plume has a pronounced elongated shape.

(B) Dispersion dominates. The plume has a relatively circular shape.

(C) Diffusion dominates. The plume has a pronounced elongated shape.

(D) Diffusion dominates. The plume has a relatively circular shape.

22. A dye tracer is used to measure the flow velocity in a 16 in sewer line. The dye is introduced into the sewer at a manhole located 1076 ft upstream of the observation manhole. The depth of flow in the upstream sewer is 3.8 in, and the depth of flow in the observation manhole is 5.2 in. The dye is first observed in the observation manhole 398 sec after it is introduced into the sewer, and the last trace of the dye is observed after 512 sec. What is the approximate average flow rate between the two manholes?

(A) 0.021 ft^3/sec

(B) 0.76 ft^3/sec

(C) 0.82 ft^3/sec

(D) 0.93 ft^3/sec

23. Characteristics of a wastewater treatment process are given.

influent five-day biochemical oxygen demand (BOD$_5$)	217 mg/L
influent ultimate biochemical oxygen demand (BOD$_u$)	312 mg/L
effluent BOD$_5$	20 mg/L
effluent total suspended solids	18 mg/L
biodegradable fraction of effluent total suspended solids	0.62

The stoichiometric oxygen demand for cell oxidation is 1.42 g/g. What is the approximate concentration of the influent soluble BOD$_5$ that escapes treatment?

(A) 4.2 mg/L

(B) 9.0 mg/L

(C) 12 mg/L

(D) 20 mg/L

24. An activated sludge plant operates with a mean cell residence time of 10 d to treat a flow of 18 925 m³/d with an influent biochemical oxygen demand (BOD) concentration of 247 mg/L. The plant wastes sludge at 34 kg/d. What is most nearly the food-to-microorganism ratio for the plant?

(A) 0.014 d^{-1}

(B) 14 d^{-1}

(C) 140 d^{-1}

(D) 1400 d^{-1}

25. An activated sludge process has the characteristics given.

influent biochemical oxygen demand (BOD) from the primary clarifier	312 mg/L
influent ammonia nitrogen from the primary clarifier	51 mg/L as N
effluent total BOD	20 mg/L
effluent ammonia nitrogen	1 mg/L as N
maximum growth rate constant	0.50 d^{-1}
maximum specific bacterial growth rate for nitrification	0.41 d^{-1}
half velocity constant	2.6 mg/L
endogenous decay rate coefficient	0.07 d^{-1}

What is most nearly the minimum required solids residence time?

(A) 2.3 d

(B) 2.6 d

(C) 3.1 d

(D) 23 d

26. Major provisions of the Clean Water Act (CWA) address all of the following EXCEPT

(A) categorical pretreatment standards for industrial effluents

(B) maximum contaminant levels for groundwater remediation

(C) national permit system for surface water discharges

(D) priority pollutants for regulation by discharge standards

27. A thickener is used to clarify a wastewater flow. The total flow to the thickener is 0.86 MGD, and the flow has an initial solids concentration of 2600 mg/L. A column settling test shows that it takes 21 min using a 6 ft settling column for thickening to occur. The slope of the upper linear portion of the settling curve is 0.23 ft/min. What is most nearly the surface area required for the thickener to produce a thickened solids concentration of 10 000 mg/L?

(A) 200 ft^2

(B) 260 ft^2

(C) 280 ft^2

(D) 350 ft^2

28. A wastewater treatment plant (WWTP) discharges raw sewage into a river during periods of high rainfall. Typical discharge flows are 15×10^6 gal/day with dissolved oxygen concentrations of 1.20 mg/L. During these periods, the river flows at 2000 ft³/sec and has a dissolved oxygen concentration of 8.10 mg/L. What is most nearly the dissolved oxygen concentration in the river once complete mixing of the WWTP discharge has occurred?

(A) 4.65 mg/L

(B) 8.02 mg/L

(C) 8.05 mg/L

(D) 8.18 mg/L

29. Which statement best defines priority pollutants as regulated under the Clean Water Act (CWA)?

(A) They are the most toxic of known chemicals.

(B) They are chemicals with relatively high toxicity and high production volume.

(C) They are chemicals that meet specific criteria of toxicity, flammability, corrosivity, or reactivity.

(D) They are chemicals associated with National Priorities List (NPL) sites.

30. Which statement best describes the relationship between assimilative capacity, stock pollutants, and fund pollutants?

(A) Assimilative capacity, being associated with fund pollutants only, is not influenced by stock pollutants.

(B) Assimilative capacity, being associated with stock pollutants only, is not influenced by fund pollutants.

(C) Assimilative capacity is relatively low for fund pollutants and relatively high for stock pollutants.

(D) Assimilative capacity is relatively high for fund pollutants and relatively low for stock pollutants.

31. Toxicity test results for a chemical are shown. Twenty fathead minnow are used for each dilution. The predilution chemical concentration was 1.63 mg/L. What is the approximate 96 h LC50 for the chemical?

concentration (%)	survivors at 96 h
2	19
4	12
8	7
16	0

(A) 0.088 mg/L

(B) 0.10 mg/L

(C) 0.30 mg/L

(D) 1.5 mg/L

32. Results of a multiple-tube fermentation test are given. What is the best approximation of the most probable number (MPN)?

volume (mL)	positives	negatives
10	5	0
1	3	2
0.1	2	3
0.01	1	4

(A) 2/100 mL

(B) 100/100 mL

(C) 140/100 mL

(D) 170/100 mL

33. The results of an ion analysis of a groundwater sample are summarized in the table. Are all ions that are likely to be present at significant concentration included in the analysis?

ion	concentration (mg/L)
Ca^{++}	128
Mg^{++}	66
SO_4^{2-}	83
Cl^-	21
NO_3^-	14
HCO_3^-	279
Na^+	7

(A) Yes, anions and cations balance.

(B) Yes, analysis is deficient in cations.

(C) No, analysis is deficient in anions.

(D) No, analysis is deficient in cations.

34. A real estate developer proposes a project that will include 346 single family houses, each located on a $1/8$ ac cleared lot. The fire demand is 750 gpm based on house separation distance, 1250 gpm based on dwelling type, and 1190 gpm based on population. The average occupancy per dwelling is four people. The daily maximum flow multiplier for domestic use during fire demand is 1.5. Most nearly, what design flow should the water line servicing the development carry if the average annual per capita demand is 165 gal/person-day?

(A) 990 gpm

(B) 1300 gpm

(C) 1400 gpm

(D) 1500 gpm

35. A rectangular tank is 2.5 m wide, 15 m long, and 3.0 m deep. It receives a flow of 900 m³/d. The hydraulic efficiency of the tank is 83%. What is most nearly the actual hydraulic detention time for the tank?

(A) 2.3 h

(B) 2.5 h

(C) 3.0 h

(D) 3.6 h

36. The design parameters for a flocculation basin are given.

design flow rate per flocculator	$18\,000$ m³/d
average time-velocity gradient	4.5×10^4, unitless
average velocity gradient	40 s⁻¹
number of sections	3
depth	3.5 m
paddle configuration	horizontal paddle wheel, axis perpendicular to flow

What are most nearly the length and width of the flocculation basin?

(A)   length $= 10.5$ m, width $= 6.5$ m

(B)   length $= 10.5$ m, width $= 10$ m

(C)   length $= 15$ m, width $= 5.0$ m

(D)   length $= 16$ m, width $= 4.0$ m

37. Water containing a volatile organic compound (VOC) is treated using air stripping. The characteristics of the VOC are given. The water temperature is 20°C, and the flow rate of the water is 37 L/s. The stripping factor is 3.5. What is most nearly the air flow rate required to treat the water?

molecular weight	112 g/mol
vapor pressure	0.26 atm at 20°C
solubility in water	7300 mg/L at 20°C

(A)   32 L/s

(B)   130 L/s

(C)   780 L/s

(D)   3600 L/s

38. Ion exchange is used to control trace metals that have contaminated a water supply source. The treatment flow rate is 675 m³/d, and the target metal is zinc, Zn^{++}, which is present at a concentration of 27 mg/L. The exchange resin has a capacity of 0.93 eq/L, and the exchanger is sized for a one-day regeneration cycle. What is most nearly the resin volume required for a one-day regeneration cycle?

(A) 0.30 m³/d

(B) 0.60 m³/d

(C) 7.6 m³/d

(D) 23 m³/d

39. Settling column results for a type II suspension are given. The percentages represent the incremental removal efficiency for the time-depth pair. What is most nearly the settling efficiency for a settling time of 90 min and a settling zone depth of 1.5 m?

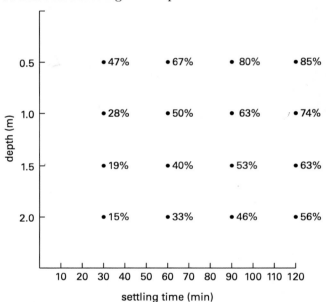

(A)   53%

(B)   65%

(C)   72%

(D)   97%

40. The solid waste characteristics for a city are given.

component	discarded mass (%)	discarded moisture (%)	discarded density (lbm/ft^3)
paper	38	6	5 —
garden	25	60	7 —
food	9	70	18 —
cardboard	8	5	3
wood	8	20	15
plastic	7	2	4
inert materials	5	8	30

The per capita solid waste generation rate for the 100,000 residents of the city is 5 lbm/person-day.

The city wants to investigate potential revenue sources from selling composted mulch composed of all of the garden and food waste and some of the paper waste. Paper should contribute 10% (by volume) of the finished mulch. The mulch would be sold at 40% moisture (by volume) for $10/yd^3.

What is the approximate annual revenue potential from the sale of the mulch?

(A) $2.0 million/yr

(B) $2.6 million/yr

(C) $2.9 million/yr

(D) $3.1 million/yr

STOP!

DO NOT CONTINUE!

This concludes the Afternoon Session of the examination. If you finish early, check your work and make sure that you have followed all instructions. After checking your answers, you may turn in your examination booklet and answer sheet and leave the examination room. Once you leave, you will not be permitted to return to work or change your answers.

Practice Exam 2 Instructions

In accordance with the rules established by your state, you may use textbooks, handbooks, bound reference materials, and any approved battery- or solar-powered, silent calculator to work this examination. However, no blank papers, writing tablets, unbound scratch paper, or loose notes are permitted. Sufficient room for scratch work is provided in the Examination Booklet.

You are not permitted to share or exchange materials with other examinees. However, the books and other resources used in this afternoon session do not have to be the same as those used in the morning session.

You will have four hours in which to work this session of the examination. Your score will be determined by the number of questions that you answer correctly. There is a total of 40 questions. All 40 questions must be worked correctly in order to receive full credit on the exam. There are no optional questions. Each question is worth 1 point. The maximum possible score for this section of the examination is 40 points.

Partial credit is not available. No credit will be given for methodology, assumptions, or work written in your Examination Booklet.

Record all of your answers on the Answer Sheet. No credit will be given for answers marked in the Examination Booklet. Mark your answers with the official examination pencil provided to you. Marks must be dark and must completely fill the bubbles. Record only one answer per question. If you mark more than one answer, you will not receive credit for the question. If you change an answer, be sure the old bubble is erased completely; incomplete erasures may be misinterpreted as answers.

If you finish early, check your work and make sure that you have followed all instructions. After checking your answers, you may turn in your Examination Booklet and Answer Sheet and leave the examination room. Once you leave, you will not be permitted to return to work or change your answers.

When permission has been given by your proctor, break the seal on the Examination Booklet. Check that all pages are present and legible. If any part of your Examination Booklet is missing, your proctor will issue you a new Booklet.

WAIT FOR PERMISSION TO BEGIN

Name: _____
      Last   First   Middle Initial

Examinee number: _____

Examination Booklet number: _____

Principles and Practice of Engineering Examination

Afternoon Session
Practice Exam 2

Practice Exam 2 Answer Sheet

Name: _____

      Last   First   Middle Initial

Date: _____

41. Ⓐ Ⓑ Ⓒ Ⓓ	51. Ⓐ Ⓑ Ⓒ Ⓓ	61. Ⓐ Ⓑ Ⓒ Ⓓ	71. Ⓐ Ⓑ Ⓒ Ⓓ
42. Ⓐ Ⓑ Ⓒ Ⓓ	52. Ⓐ Ⓑ Ⓒ Ⓓ	62. Ⓐ Ⓑ Ⓒ Ⓓ	72. Ⓐ Ⓑ Ⓒ Ⓓ
43. Ⓐ Ⓑ Ⓒ Ⓓ	53. Ⓐ Ⓑ Ⓒ Ⓓ	63. Ⓐ Ⓑ Ⓒ Ⓓ	73. Ⓐ Ⓑ Ⓒ Ⓓ
44. Ⓐ Ⓑ Ⓒ Ⓓ	54. Ⓐ Ⓑ Ⓒ Ⓓ	64. Ⓐ Ⓑ Ⓒ Ⓓ	74. Ⓐ Ⓑ Ⓒ Ⓓ
45. Ⓐ Ⓑ Ⓒ Ⓓ	55. Ⓐ Ⓑ Ⓒ Ⓓ	65. Ⓐ Ⓑ Ⓒ Ⓓ	75. Ⓐ Ⓑ Ⓒ Ⓓ
46. Ⓐ Ⓑ Ⓒ Ⓓ	56. Ⓐ Ⓑ Ⓒ Ⓓ	66. Ⓐ Ⓑ Ⓒ Ⓓ	76. Ⓐ Ⓑ Ⓒ Ⓓ
47. Ⓐ Ⓑ Ⓒ Ⓓ	57. Ⓐ Ⓑ Ⓒ Ⓓ	67. Ⓐ Ⓑ Ⓒ Ⓓ	77. Ⓐ Ⓑ Ⓒ Ⓓ
48. Ⓐ Ⓑ Ⓒ Ⓓ	58. Ⓐ Ⓑ Ⓒ Ⓓ	68. Ⓐ Ⓑ Ⓒ Ⓓ	78. Ⓐ Ⓑ Ⓒ Ⓓ
49. Ⓐ Ⓑ Ⓒ Ⓓ	59. Ⓐ Ⓑ Ⓒ Ⓓ	69. Ⓐ Ⓑ Ⓒ Ⓓ	79. Ⓐ Ⓑ Ⓒ Ⓓ
50. Ⓐ Ⓑ Ⓒ Ⓓ	60. Ⓐ Ⓑ Ⓒ Ⓓ	70. Ⓐ Ⓑ Ⓒ Ⓓ	80. Ⓐ Ⓑ Ⓒ Ⓓ

Practice Exam 2

41. A gate that is hinged at the top is used to prevent backflow of tidal water into a 0.5 m circular stormwater drain line. At high tide, the hinge at the top of the gate is 2.3 m below the water surface. What force is most nearly required to open the gate at high tide?

(A) 3.0 kN

(B) 3.8 kN

(C) 12 kN

(D) 19 kN

42. 20°C water is pumped from a tank through a pipe to a point 340 m above the tank. The pump is located at the same elevation as the tank's water surface. The pipe's inside diameter is 2.54 cm for most of its length, but decreases to a nozzle with an inside diameter of 0.5 cm. The water discharges from the nozzle at 7000 kPa. The flow in the pipe is 0.002 m³/s. What approximate pressure is required at the discharge side of the pump?

(A) 1600 kPa

(B) 7500 kPa

(C) 10 000 kPa

(D) 15 000 kPa

43. Three water storage tanks are connected at a common node C as shown. The flow between node C and tank D is 4 ft³/sec. The tank elevations, pipe lengths, and nominal diameters are shown.

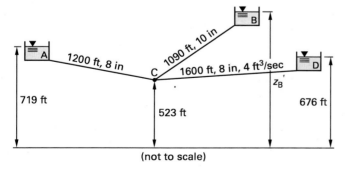

(not to scale)

The pipe cross-sectional areas and corresponding Darcy friction factors are as shown.

pipe	nominal diameter (in)	cross-sectional area (ft²)	friction factor, f
AC	8	0.35	0.016
CB	10	0.55	0.014
DC	8	0.35	0.016

What is most nearly the water surface elevation in tank B?

(A) 420 ft

(B) 510 ft

(C) 540 ft

(D) 550 ft

44. A pneumatically operated valve at the end of 30 m of 200 mm schedule-40 steel pipe closes abruptly. The pipe is anchored against any axial movement. With the valve fully open, the water velocity in the pipe is 2.4 m/s. The modulus of elasticity of the steel pipe is 200×10^9 N/m², and the bulk modulus of elasticity is 2.07×10^9 N/m². What is most nearly the maximum head produced by the valve closure?

(A) 9.8 m

(B) 310 m

(C) 330 m

(D) 590 m

45. A pipe friction apparatus has two sets of water manometers. The first set of manometers is separated by 10 cm of straight pipe and includes the pipe only, with no fittings or valves. The second set of manometers includes a fully open gate valve with 20 cm of straight pipe at either end. The manometer readings for a water velocity of 3.2 m/s are shown. What is most nearly the loss coefficient for a fully open gate valve placed between the second set of manometers?

manometer	reading (cm)
upstream, no valve	14.2
downstream, no valve	12.9
upstream, with valve	18
downstream, with valve	3.7

(A) 0.17

(B) 0.22

(C) 0.25

(D) 0.27

46. A 1000 ft long, 10 in cement-lined ductile cast iron pipe is used in a water transmission line to deliver a flow of 950 gpm. The water temperature is 70°F. The design specific roughness for cement-lined cast iron pipe is 0.000008 ft. What is most nearly the head loss in the pipe?

(A) 0.50 ft

(B) 1.8 ft

(C) 4.0 ft

(D) 6.1 ft

47. What is the approximate flow rate through a 24 in wide Cipoletti weir when the water height above the notch is 8.6 in?

(A) 3.8 ft³/sec

(B) 4.1 ft³/sec

(C) 5.4 ft³/sec

(D) 14 ft³/sec

48. A trapezoidal channel has concrete walls placed on a 1:1 slope and a gravel bottom. The channel slope is 0.0005 m/m, and the normal water depth is 1 m. Because of the gravel bottom, the maximum flow velocity in the channel is limited to 0.75 m/s. The width of the channel base is most nearly

(A) 0.62 m

(B) 0.87 m

(C) 1.8 m

(D) 3.6 m

49. Two reservoirs are connected by a 9 km long earth-lined channel with a constant slope. The difference in elevation between the reservoirs is 20 m. The channel bottom is 2 m wide with a normal water depth of 1 m,

and the channel has 3:1 horizontal-to-vertical sides. What is most nearly the flow rate of water in the channel?

(A) 4.2 m³/s

(B) 9.3 m³/s

(C) 37 m³/s

(D) 49 m³/s

50. The low-head siphon spillway shown is used to divert irrigation flow from a canal. The characteristics of the siphon spillway are given in the table.

discharge flow	90 ft³/sec
operating head, h	4 ft
atmospheric pressure head	34 ft
radius to throat centerline, R_c	2 ft
siphon entrance coefficient	0.6

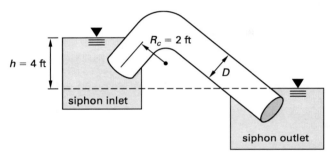

What is the approximate diameter, D, required for the siphon throat?

(A) 0.86 ft

(B) 1.2 ft

(C) 2.0 ft

(D) 3.5 ft

51. The culvert inlet and outlet shown are square flush and constructed of reinforced concrete. The culvert barrel is round reinforced concrete pipe and has a slope of 83:1.

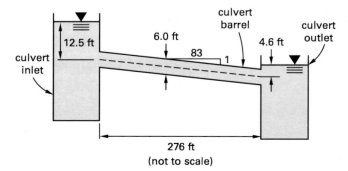

The maximum discharge under the conditions shown is most nearly

(A) 360 ft³/sec

(B) 410 ft³/sec

(C) 470 ft³/sec

(D) 560 ft³/sec

52. A circular channel flowing half full has a hydraulic radius of 1 m. The channel has a Manning roughness coefficient of 0.013. What is most nearly the equivalent Darcy friction factor?

(A) 0.0060

(B) 0.013

(C) 0.023

(D) 0.029

53. A unit hydrograph for a drainage area is shown. Twelve hours after the beginning of runoff, the discharge measured at a gaging station at the outlet of the drainage area is 150 m³/s. What is the approximate peak discharge from the drainage area?

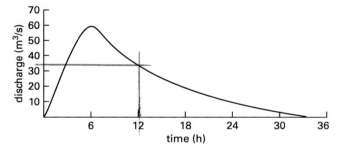

(A) 59 m³/s

(B) 180 m³/s

(C) 210 m³/s

(D) 260 m³/s

54. An isohyetal map for a defined region is shown. The gross area enclosed by each isohyet within the region boundary is summarized in the table. What is most nearly the areal average precipitation for the region?

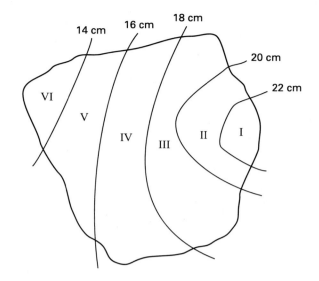

area	isohyet (cm)	enclosed area (km²)
I	> 22	84
II	20	252
III	18	578
IV	16	892
V	14	1136
VI	< 14	1294

(A) 15 cm

(B) 17 cm

(C) 18 cm

(D) 52 cm

55. A histograph of peak flows for a 112 yr period is shown.

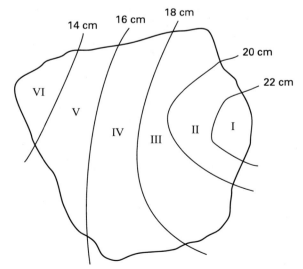

What is the approximate recurrence interval for the event corresponding to a flow of 40,000 ft³/sec?

(A) 0.35 yr

(B) 2.2 yr

(C) 2.8 yr

(D) 3.4 yr

56. A small, developed area covered with manicured sod and well-defined drainage channels has a flow distance of 270 ft and an average surface slope of 0.011 ft/ft. The rainfall intensity for the area from the 15 min duration, 10 yr storm is 2.1 in/hr. What is most nearly the time of concentration?

(A) 2.7 min

(B) 26 min

(C) 29 min

(D) 38 min

57. A stormwater detention pond uses a submerged orifice to control discharge to an open channel. A standpipe is used to prevent the pool elevation from exceeding 100 ft. The centerline of the orifice is at an elevation of 92.6 ft, and the orifice opening is sharp edged. The diameter of the orifice required to limit the orifice discharge to 20 ft³/sec is most nearly

(A) 0.77 ft

(B) 0.85 ft

(C) 1.1 ft

(D) 1.4 ft

58. Annual flood data compiled for a 97 yr period are given.

- The arithmetic mean of the log value of all floods (for flow units of ft³/sec) is 3.571.

- The sum of the squared difference of the log of the flood magnitude for each probability, and the arithmetic mean log value of all floods is 3.894.

- The sum of the cubed difference of the log of the flood magnitude for each probability, and the arithmetic mean of the log value of all floods is 0.181.

What is most nearly the flow magnitude of the 25 yr flood?

(A) 4100 ft³/sec

(B) 8300 ft³/sec

(C) 8700 ft³/sec

(D) 10,000 ft³/sec

59. Leachate is expected to accumulate on the upper surface of a landfill that is constructed with a natural clay stratum as the liner. The clay stratum has a diffusion coefficient of 8.7×10^{-9} m²/s and a tortuosity of 0.6. The chloride concentration in the leachate against the upper surface of the clay is expected to reach 12 000 mg/L. Chloride is used as the tracer. Approximately how thick should the clay stratum be to prevent the chloride concentration from reaching 100 mg/L on the underside of the liner until 100 yr have passed?

(A) 8.0 m

(B) 15 m

(C) 20 m

(D) 61 m

60. Flow lines for a contaminated groundwater extraction system are shown. Which circled letter on the illustration identifies a stagnation zone in the extraction system?

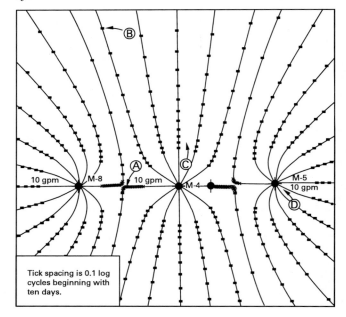

Tick spacing is 0.1 log cycles beginning with ten days.

(A) A

(B) B

(C) C

(D) D

61. Two 8 in diameter auger holes are drilled 10 ft apart through a 15 ft thick shallow unconfined aquifer into an underlying clay layer. Water is pumped from one hole to the other through a flow meter at 26 gpm until the water level in each hole stabilizes with a head difference between the holes of 9.7 in. The auger holes are of equal depth. What is most nearly the hydraulic conductivity of the soil comprising the aquifer?

(A) 30 ft/day

(B) 130 ft/day

(C) 360 ft/day

(D) 620 ft/day

62. What is most nearly the required width of a rectangular horizontal-flow grit chamber where the conditions given apply?

flow rate	3.5×10^6 gal/day
depth	4 ft
mean particle diameter	0.22 mm
grit specific gravity	2.65
Camp formula constant	0.05
Darcy friction factor	0.03

(A) 0.60 ft

(B) 1.9 ft

(C) 8.8 ft

(D) 14 ft

63. A mixed liquor suspended solids (MLSS) suspension has an initial concentration of 2400 mg/L that settles to the 356 mL mark in a 1 L graduated cylinder after 30 min. What is most nearly the sludge volume index of the suspension?

(A) 150 mL/g

(B) 270 mL/g

(C) 3700 mL/g

(D) 6700 mL/g

64. A municipality has selected an activated sludge process to provide a denitrified effluent. Selected data relevant to denitrification of the municipality's wastewater are given.

growth rate	0.38 d^{-1}
yield coefficient	0.81 g/g
wastewater temperature	16°C
methanol concentration	72 mg/L
influent nitrate concentration	29 mg/L
influent nitrite concentration	8 mg/L
half-velocity constant for methanol	12 mg/L
half-velocity constant for nitrogen	0.31 mg/L
temperature correction coefficient	1.1

What is most nearly the corrected maximum growth rate for denitrification?

(A) 0.011 d^{-1}

(B) 0.12 d^{-1}

(C) 0.17 d^{-1}

(D) 0.21 d^{-1}

65. A constructed wetland for wastewater treatment (CWWT) is proposed for a community of 800 people. The CWWT design criteria imposed by the local health department are 6.0 m²/PE and 0.2 m³/PE·d with a minimum bed depth of 0.75 m. What is most nearly the CWWT's empty-bed hydraulic residence time?

(A) 0.028 d

(B) 0.044 d

(C) 23 d

(D) 40 d

66. A wastewater treatment plant experienced an emergency bypass of 4.2 million gallons of untreated domestic wastewater into a freshwater stream. If the bacteria die at a rate of 1.0 d⁻¹, approximately how much time is required for 90% of the bacteria in the initial release to die?

(A) 0.11 d

(B) 0.69 d

(C) 2.3 d

(D) 10 d

67. A three-stage rotating biological contactor (RBC) is used to treat a wastewater flow of 0.5 MGD. The wastewater has an influent soluble BOD_5 of 176 mg/L and an effluent BOD_5 of 40 mg/L. Pilot studies give the RBC a design loading rate of 40.6 $L/m^2 \cdot d$. What is most nearly the total disc surface area of the RBC?

(A) 30 000 m^2

(B) 55 000 m^2

(C) 89 000 m^2

(D) 160 000 m^2

68. What is most nearly the volatile suspended solids (VSS) concentration of the following wastewater sample?

sample volume filtered (VF)	200 mL
sample volume evaporated (VD)	100 mL
mass of dried crucible and filter paper (MS)	25.439 g
mass of dry evaporation dish (MD)	275.41 g
mass of dried crucible, filter paper, and solids (MSS)	25.645 g
mass of dried evaporation dish and solids (MDS)	276.227 g
mass of ignited crucible, filter paper, and solids (MSI)	25.501 g
mass of ignited evaporation dish and solids (MDI)	276.201 g

(A) 260 mg/L

(B) 310 mg/L

(C) 720 mg/L

(D) 1000 mg/L

69. The coliform group is used as indicator organisms for pathogens because they have all of the following characteristics EXCEPT they

(A) apply to all types of water

(B) are always present when pathogens are present and normally absent otherwise ✓

(C) are easily detectable by routine analytical methods ✓

(D) are themselves pathogenic

70. A well is contaminated with 100 μg/L trichloroethylene (TCE) and 7.2 μg/L 1,1-dichloroethylene (1,1-DCE). The potency factor is 0.011 $(mg/kg \cdot d)^{-1}$ for TCE and 0.58 $(mg/kg \cdot d)^{-1}$ for 1,1-DCE. What is the approximate lifetime risk to an adult from ingesting drinking water from the contaminated well?

(A) 32 in one million

(B) 64 in one million

(C) 76 in one million

(D) 770 in one million

71. An industrial plant proposes to return noncontact cooling water to a river. The river water temperature upstream of the mixing zone is 6°C, and the returned cooling water temperature is 20°C. The river flows at 280 m^3/s, and the cooling water flow is 11 m^3/s. The reaeration and deoxygenation river constants are equal at 0.080 d^{-1}. What is the approximate deoxygenation rate constant downstream of the mixing zone?

(A) 0.075 d^{-1}

(B) 0.085 d^{-1}

(C) 0.19 d^{-1}

(D) 0.47 d^{-1}

72. Biochemical oxygen demand (BOD) and theoretical oxygen demand (ThOD) for selected chemical wastes are given.

	BOD (g/g)	ThOD (g/g)	
chemical waste A	2.15	2.52	0.37
chemical waste B	1.34	2.15	0.81
chemical waste C	1.85	1.92	0.07
chemical waste D	1.64	2.91	1.27

Which chemical waste listed in the table is the most likely to be biologically degradable?

(A) chemical waste A, because it has the greatest BOD

(B) chemical waste B, because it has the least BOD

(C) chemical waste C, because it has the greatest BOD to ThOD ratio

(D) chemical waste D, because it has the smallest BOD to ThOD ratio

73. Surface water runoff with an average ammonia nitrogen concentration of 1.2 mg/L contributes 735,000 ft^3 of water per year to a lake with a constant volume of 1.9×10^6 ft^3. The ammonia nitrogen concentration is 0.26 mg/L in the stream flowing out of the lake. All ammonia depletion in the lake is from nitrification. What is most nearly the oxygen demand exerted on the lake from the ammonia nitrogen?

(A)   25 kg O_2/yr

(B)   46 kg O_2/yr

(C)   91 kg O_2/yr

(D)   230 kg O_2/yr

74. A mass diagram for a municipal water storage tank is shown.

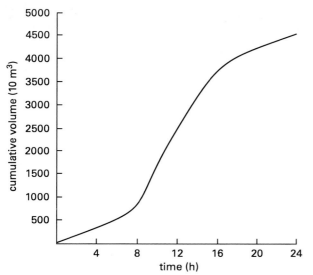

What is most nearly the minimum required capacity of the storage tank?

(A)   1300 m^3

(B)   6500 m^3

(C)   13 000 m^3

(D)   45 000 m^3

75. Three continuous-flow stirred tank reactors (CSTRs) in series are used in a chemical treatment process to control taste and odor in drinking water. What is most nearly the residence time for 90% degradation at a reaction rate of 0.37 h^{-1}?

(A)   0.97 h

(B)   2.4 h

(C)   3.2 h

(D)   7.3 h

76. The chlorination curve shown was prepared from samples of a treated drinking water.

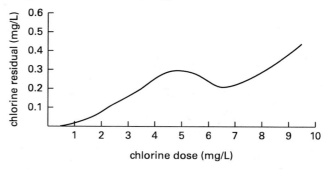

What is most nearly the chlorine dose needed to produce a free chlorine residual of 0.2 mg/L?

(A)   3.5 mg/L

(B)   4.8 mg/L

(C)   6.6 mg/L

(D)   9.2 mg/L

77. What is most nearly the Langelier stability index of the following water sample?

pH	7.4
Ca^{++}	46 mg/L
HCO_3^-	133 mg/L
ionic strength and temperature constant	2.28

(A)   −0.60

(B)   −0.48

(C)   0.15

(D)   0.43

78. An impeller-type flash mixer treats 20 000 m^3/d of 10°C water. The mixer has a minimum velocity gradient of 960 s^{-1} with a 90 s contact period. What is most nearly the power requirement for the mixer?

(A)   3.4 kW

(B)   25 kW

(C)   34 kW

(D)   250 kW

79. A sample of hard water is analyzed, and the results are tabulated as given. The flow rate of the hard water is 12 000 m^3/d. The water is to be softened to a hardness of 65 mg/L as $CaCO_3$, including residual hardness. No extra sodium hydroxide is required for pH adjustment. Determine the approximate amount of sodium hydroxide required per day to soften the water.

ion	C (mg/L)	C_{molar} (mmol/L)
Ca^{++}	112	2.8
Mg^{++}	46	1.9
HCO_3^-	348	5.7

(A) 2200 kg/d

(B) 4600 kg/d

(C) 5500 kg/d

(D) 5600 kg/d

80. Two site alternatives are proposed for a municipal solid waste landfill. The sites have been rated against criteria on a weighted scale from 1 (unimportant/poor) to 4 (very important/excellent) as given in the table. What is most nearly the weighted average for the most desirable site?

category	criteria	weighting factor	site 1 rating	site 2 rating
location	haul distance	2	2	3
	access routes	4	3	1
	land value	3	2	2
soil/geology	permeability	3	3	4
	heterogeneities	4	2	3
	cover quantities	3	3	2
	seismic activity	4	4	4
groundwater	quality	2	3	3
	gradient	1	2	3
	depth	2	4	3
hydrology	drainage pattern	3	3	2
	streams	2	3	2
community	population	3	2	3
	land uses	4	4	3
	opposition	4	3	2

44   129   116

(A) 0.98

(B) 2.6

(C) 2.9

(D) 3.0

STOP!

DO NOT CONTINUE!

This concludes the Afternoon Session of the examination. If you finish early, check your work and make sure that you have followed all instructions. After checking your answers, you may turn in your examination booklet and answer sheet and leave the examination room. Once you leave, you will not be permitted to return to work or change your answers.

Answer Keys

Practice Exam 1 Answer Key

1. C	11. D	21. D	31. A
2. A	12. B	22. B	32. D
3. B	13. D	23. B	33. C
4. B	14. B	24. B	34. D
5. A	15. D	25. C	35. B
6. C	16. D	26. B	36. A
7. B	17. B	27. C	37. C
8. B	18. B	28. B	38. B
9. C	19. C	29. B	39. C
10. B	20. B	30. D	40. A

Practice Exam 2 Answer Key

41. A	51. B	61. C	71. B
42. D	52. B	62. B	72. C
43. B	53. D	63. A	73. C
44. B	54. C	64. C	74. C
45. A	55. C	65. C	75. C
46. C	56. D	66. C	76. D
47. B	57. D	67. C	77. B
48. A	58. C	68. C	78. B
49. B	59. B	69. D	79. B
50. D	60. A	70. B	80. C

Solutions

Practice Exam 1

1. Apply the energy equation.

g	gravitational acceleration, 9.81	m/s^2
p	pressure	kN/m^2
v	flow velocity	m/s
z	elevation	m
ρ	water density	kg/m^3

$$\frac{p_A}{\rho g} + z_A + \frac{v_A^2}{2g} = \frac{p_B}{\rho g} + z_B + \frac{v_B^2}{2g} = \frac{p_C}{\rho g} + z_C + \frac{v_C^2}{2g}$$

$$p_A = 0 \ kN/m^2 \quad \text{[atmospheric]}$$
$$p_C = 0 \ kN/m^2 \quad \text{[free jet]}$$
$$z_A = 100 \ m$$

$$\frac{p_C}{\rho g} + z_C + \frac{v_C^2}{2g} = 0 + 34 \ m + \frac{v_C^2}{(2)\left(9.81 \ \dfrac{m}{s^2}\right)}$$

$$v_C = 36 \ m/s$$

A	pipe cross-sectional area	m^2
D	pipe diameter	m

$$A_C = \frac{\pi D_C^2}{4} = \frac{\pi\left(\dfrac{8 \ cm}{100 \ \dfrac{cm}{m}}\right)^2}{4} = 0.0050 \ m^2$$

Q	flow rate	m^3/s

$$Q = A_C v_C = \left(0.0050 \ m^2\right)\left(36 \ \frac{m}{s}\right) = 0.18 \ m^3/s$$

$$A_B = \frac{\pi D_B^2}{4} = \frac{\pi\left(\dfrac{10 \ cm}{100 \ \dfrac{cm}{m}}\right)^2}{4} = 0.0079 \ m^2$$

$$v_B = \frac{Q}{A_B} = \frac{0.18 \ \dfrac{m^3}{s}}{0.0079 \ m^2} = 23 \ m/s$$

The density of water is $1000 \ kg/m^3$.

$$\frac{p_B}{\rho g} + z_B + \frac{v_B^2}{2g} = z_A$$

$$\frac{p_B\left(1000 \ \dfrac{N}{kN}\right)}{\left(1000 \ \dfrac{kg}{m^3}\right)\left(9.81 \ \dfrac{m}{s^2}\right)}$$

$$\times \left(1 \ \frac{N \cdot s^2}{kg \cdot m}\right)$$

$$+ \ 46 \ m + \frac{\left(23 \ \dfrac{m}{s}\right)^2}{(2)\left(9.81 \ \dfrac{m}{s^2}\right)} = 100 \ m$$

$$p_B = 265 \ kN/m^2$$
$$\left(260 \ kN/m^2\right)$$

The answer is (C).

Author Commentary

💣 The calculations in this problem are simple but involve several terms. The most common error made in problems requiring multiple conversions is an incorrect conversion of the units. Be careful to use consistent units, using meters for all length, area, and volume calculations.

2. Find the mass flow rate. The density of water is $62.4 \, \text{lbm/ft}^3$.

$\dot{m}$	mass flow rate	lbm/sec
Q	volumetric flow rate	ft³/sec
ρ	water density	lbm/ft³

$$\dot{m} = Q\rho$$
$$= \left(0.55 \, \frac{\text{ft}^3}{\text{sec}}\right)\left(62.4 \, \frac{\text{lbm}}{\text{ft}^3}\right)$$
$$= 34.3 \, \text{lbm/sec}$$

Find the pipe area.

A	pipe cross-sectional area	ft²
D	inside pipe diameter	ft

$$A = \frac{\pi D^2}{4} = \frac{\pi \left(\dfrac{3 \text{ in}}{12 \, \frac{\text{in}}{\text{ft}}}\right)^2}{4} = 0.049 \, \text{ft}^2$$

v	flow velocity	ft/sec

$$\text{v} = \frac{Q}{A} = \frac{0.55 \, \dfrac{\text{ft}^3}{\text{sec}}}{0.049 \, \text{ft}^2} = 11.2 \, \text{ft/sec}$$

Because the elbows have a constant diameter, the area is constant.

F_x	force component in the x-direction	lbf
g_c	gravitational constant, 32.2	lbm-ft/lbf-sec²
p	pressure	lbf/ft²
θ	pipe bend angle	deg

$$F_x = pA(\cos\theta - 1) + \frac{\dot{m}\text{v}(\cos\theta - 1)}{g_c}$$
$$= \left(50 \, \frac{\text{lbf}}{\text{in}^2}\right)\left(12 \, \frac{\text{in}}{\text{ft}}\right)^2 (0.049 \, \text{ft}^2)(\cos 90° - 1)$$
$$+ \frac{\left(34.3 \, \dfrac{\text{lbm}}{\text{sec}}\right)\left(11.2 \, \dfrac{\text{ft}}{\text{sec}}\right)(\cos 90° - 1)}{32.2 \, \dfrac{\text{lbm-ft}}{\text{lbf-sec}^2}}$$
$$= -364.73 \, \text{lbf}$$

F_y	force component in the y-direction	lbf

$$F_y = \left(pA + \frac{\dot{m}\text{v}}{g_c}\right)\sin\theta$$
$$= \left(\begin{array}{c} \left(50 \, \dfrac{\text{lbf}}{\text{in}^2}\right)\left(12 \, \dfrac{\text{in}}{\text{ft}}\right)^2 (0.049 \, \text{ft}^2) \\[2mm] + \dfrac{\left(34.3 \, \dfrac{\text{lbm}}{\text{sec}}\right)\left(11.2 \, \dfrac{\text{ft}}{\text{sec}}\right)}{32.2 \, \dfrac{\text{lbm-ft}}{\text{lbf-sec}^2}} \end{array}\right)\sin 90°$$
$$= 364.73 \, \text{lbf}$$

F_R	resultant force	lbf

$$F_R = \sqrt{F_x^2 + F_y^2} = \sqrt{(-364.73 \, \text{lbf})^2 + (364.73 \, \text{lbf})^2}$$
$$= 516 \, \text{lbf}$$

The answer is (A).

Author Commentary

🕐 In this problem, the mass flow rate, velocity, and area are constant, so calculate them once at the beginning, and use the same values throughout the solution.

💣 The solution requires using the mass flow rate, not the volumetric flow rate. Multiply the volumetric flow rate in ft³/sec by the water density in lbm/ft³ to find the mass flow rate.

3. Combine the pump curves by adding the heads for series operation on a graph that includes the operating curves. On the same graph, plot the combined head losses and plot the total head loss given for each flow rate. For convenience, 50 gpm, 60 gpm, and 70 gpm may be used for the flow rates given the provided range. The total head loss at these flow rates is determined by interpolation using the total head losses given. Tabulate the results.

flow rate (gpm)	head loss (ft)			
	pump 1	pump 2	combined pumps	total head loss (ft)
50	45	74	119	95.2
60	34	67	101	99.6
70	21	58	79	104.0

The intersection of the total head loss curve and the combined pump curve defines the operating point for the combined pumps. The operating point is the discharge of the combined pumps. From the illustration shown,

Q discharge of the combined pumps   gpm

$$Q = 60 \text{ gpm}$$

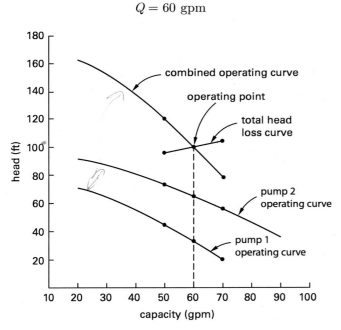

The answer is (B).

Author Commentary

🕐 The pump curves need to be added only near the point corresponding to the total head loss. Don't waste time plotting the curve outside of this range.

4. From steel pipe dimension tables, the inside cross-sectional area, A, of a 12 in diameter schedule-40 welded steel pipe is approximately 0.785 ft^2.

A	pipe cross-sectional area	ft^2
Q	flow rate	ft^3/sec
v	velocity	ft/sec

$$\text{v}_1 = \frac{Q_1}{A} = \frac{2.2 \, \dfrac{\text{ft}^3}{\text{sec}}}{0.785 \text{ ft}^2} = 2.8 \text{ ft/sec}$$

$$\text{v}_2 = \frac{Q_2}{A} = \frac{4.4 \, \dfrac{\text{ft}^3}{\text{sec}}}{0.785 \text{ ft}^2} = 5.6 \text{ ft/sec}$$

From water properties tables, the kinematic viscosity at 70°F is

ν	kinematic viscosity	ft^2/sec

$$\nu = 1.059 \times 10^{-5} \text{ ft}^2/\text{sec}$$

D	diameter	ft
Re	Reynolds number	—

$$\text{Re}_1 = \frac{D\text{v}_1}{\nu} = \frac{\left(\dfrac{12 \text{ in}}{12 \, \frac{\text{in}}{\text{ft}}}\right)\left(2.8 \, \dfrac{\text{ft}}{\text{sec}}\right)}{1.059 \times 10^{-5} \, \dfrac{\text{ft}^2}{\text{sec}}} = 2.6 \times 10^5$$

$$\text{Re}_2 = \frac{D\text{v}_2}{\nu} = \frac{\left(\dfrac{12 \text{ in}}{12 \, \frac{\text{in}}{\text{ft}}}\right)\left(5.6 \, \dfrac{\text{ft}}{\text{sec}}\right)}{1.059 \times 10^{-5} \, \dfrac{\text{ft}^2}{\text{sec}}} = 5.3 \times 10^5$$

Using a relative roughness of 0.0002 ft, from the Moody diagram, the Darcy friction factors are

f Darcy friction factor   –

$$f_1 = 0.0168$$
$$f_2 = 0.0152$$

Rearranging the equation for head loss, determine the loss coefficient, k, for each extreme.

g	gravitational acceleration, 32.2	ft/sec^2
h_f	head loss	ft
k	loss coefficient	–
L	length	ft
n	loss coefficient	–

$$h_f = kQ^n = \frac{fL\mathrm{v}^2}{2Dg}$$

$$k_1 = \frac{f_1 L \mathrm{v}_1^2}{2DgQ_1^n} = \frac{(0.0168)(1\ \text{mi})\left(5280\ \dfrac{\text{ft}}{\text{mi}}\right)\left(2.8\ \dfrac{\text{ft}}{\text{sec}}\right)^2}{(2)\left(\dfrac{12\ \text{in}}{12\ \dfrac{\text{in}}{\text{ft}}}\right)\left(32.2\ \dfrac{\text{ft}}{\text{sec}^2}\right)\left(2.2\ \dfrac{\text{ft}^3}{\text{sec}}\right)^n}$$

$$= \frac{10.8}{2.2^n}$$

$$k_2 = \frac{f_2 L \mathrm{v}_2^2}{2DgQ_2^n} = \frac{(0.0152)(1\ \text{mi})\left(5280\ \dfrac{\text{ft}}{\text{mi}}\right)\left(5.6\ \dfrac{\text{ft}}{\text{sec}}\right)^2}{(2)\left(\dfrac{12\ \text{in}}{12\ \dfrac{\text{in}}{\text{ft}}}\right)\left(32.2\ \dfrac{\text{ft}}{\text{sec}^2}\right)\left(4.4\ \dfrac{\text{ft}^3}{\text{sec}}\right)^n}$$

$$= \frac{39.1}{4.4^n}$$

The loss coefficients for all flows are equivalent. The loss coefficient of the pipe is

$$k_1 = k_2$$
$$\frac{10.8}{2.2^n} = \frac{39.1}{4.4^n}$$
$$\left(\frac{4.4}{2.2}\right)^n = \frac{39.1}{10.8}$$
$$\log 2^n = \log 3.62$$
$$n = 1.86$$
$$k = \frac{10.8}{2.2^n} = \frac{10.8}{(2.2)^{1.86}} = 2.49 \quad (2.5)$$

The answer is (B).

Author Commentary

🕐 The loss coefficient, k, must be found twice, once for each of the two extreme flow rates given in the problem statement. This will require calculating the velocity, the Reynolds number, and k (in terms of n) twice. To save time, perform each calculation in pairs using each of the flow rates, and set the two values for k (in terms of n) equal to each other to find the actual value of k.

💣※ In the final calculation, the resulting values for each calculated k are set equal to each other and solved for n. This requires using log functions and is the only complicated calculation in the series. Be careful.

5. Find the head loss in the pipe.

h_f	head loss in the pipe	ft
z_1	exit manometer reading	ft
z_2	entrance manometer reading	ft

$$h_f = z_2 - z_1 = \frac{37.1\ \text{in} - 9.2\ \text{in}}{12\ \dfrac{\text{in}}{\text{ft}}} = 2.33\ \text{ft}$$

A	pipe cross-sectional area	ft^2
D	pipe inside diameter	ft

$$A = \frac{\pi D^2}{4} = \frac{\pi\left(\dfrac{0.75\ \text{in}}{12\ \dfrac{\text{in}}{\text{ft}}}\right)^2}{4} = 0.0031\ \text{ft}^2$$

Q	flow rate	ft^3/sec
v	flow velocity	ft/sec

$$v = \frac{Q}{A} = \frac{8 \dfrac{\text{gal}}{\text{min}}}{(0.0031 \text{ ft}^2)\left(60 \dfrac{\text{sec}}{\text{min}}\right)\left(7.48 \dfrac{\text{gal}}{\text{ft}^3}\right)} = 5.8 \text{ ft/sec}$$

f	friction factor	–
g	gravitational acceleration, 32.2	ft/sec^2
L	pipe length	ft

$$f = \frac{2h_f D g}{L v^2} = \frac{(2)(2.33 \text{ ft})\left(\dfrac{0.75 \text{ in}}{12 \dfrac{\text{in}}{\text{ft}}}\right)\left(32.2 \dfrac{\text{ft}}{\text{sec}^2}\right)}{(20 \text{ ft})\left(5.8 \dfrac{\text{ft}}{\text{sec}}\right)^2}$$

$$= 0.014$$

Assume flow is turbulent. Use the Nikuradse smooth pipe equation to calculate the pipe friction factor.

ϵ	specific roughness	ft
ϵ/D	relative roughness	–

$$\frac{1}{\sqrt{f}} = 1.74 - 2 \log\left(\frac{2\epsilon}{D}\right)$$

$$\frac{1}{\sqrt{0.014}} = 1.74 - 2 \log\left(\frac{2\epsilon}{D}\right)$$

$$\frac{\epsilon}{D} = \frac{10^{-3.36}}{2} = 0.00022$$

$$\epsilon = \left(\frac{\epsilon}{D}\right)$$

$$D = (0.00022)\left(\frac{0.75 \text{ in}}{12 \dfrac{\text{in}}{\text{ft}}}\right)$$

$$= 0.000014 \text{ ft}$$

The answer is (A).

Author Commentary

🕐 Flow rate is given, so use the friction factor instead of separately calculating area and velocity for the calculation of the friction factor. This saves a calculation step.

💣 Assume the flow is turbulent, which is reasonable for the flow rate and pipe size given. Turbulent flow suggests using the Nikuradse equation, which allows for a simpler calculation than alternative equations.

6. Rearrange the Darcy equation to solve for the diameter. Solve iteratively until the assumed and calculated friction factor values converge.

D	diameter	m
f	Darcy friction factor	–
g	gravitational acceleration, 9.81	m/s^2
h_f	head loss due to friction	m
L	length	m
Q	flow rate	m^3/s

$$h_f = \frac{f L v^2}{2Dg}$$

$$= \frac{8fLQ^2}{\pi^2 g D^5}$$

$$D = \left(\frac{8fLQ^2}{\pi^2 g h_f}\right)^{1/5}$$

$$= \left(\frac{(8)(0.015)(320 \text{ m})\left(\left(20 \dfrac{\text{L}}{\text{s}}\right)\left(0.001 \dfrac{\text{m}^3}{\text{L}}\right)\right)^2}{\pi^2\left(9.81 \dfrac{\text{m}}{\text{s}^2}\right)(1.0 \text{ m})}\right)^{1/5}$$

$$= 0.174 \text{ m}$$

Converting to inches,

$$(0.174 \text{ m})\left(\frac{1000 \dfrac{\text{mm}}{\text{m}}}{25.4 \dfrac{\text{mm}}{\text{in}}}\right) = 6.9 \text{ in}$$

Therefore, use an 8 in standard pipe size. (8 in pipe has an inside diameter of 7.942 in.) From water property tables, the kinematic viscosity of 20°C water is 1.0×10^{-6} m^2/s.

Re	Reynolds number	–
ν	kinematic viscosity	m^2/s

$$\text{Re} = \frac{4Q}{\pi D \nu} = \frac{(4)\left(20\ \dfrac{\text{L}}{\text{s}}\right)\left(0.001\ \dfrac{\text{m}^3}{\text{L}}\right)}{\pi(0.20\ \text{m})\left(1.0 \times 10^{-6}\ \dfrac{\text{m}^2}{\text{s}}\right)} = 1.27 \times 10^5$$

The specific roughness of PVC pipe is 0.0000015 m.

ϵ	specific roughness	m
ϵ/D	relative roughness	–

$$\frac{\epsilon}{D} = \frac{0.0000015\ \text{m}}{0.20\ \text{m}} = 0.0000075$$

From the Moody friction factor chart, for a Reynolds number of 1.27×10^5 and a relative roughness of 0.0000075, the Darcy friction factor is approximately 0.018.

The assumed Darcy friction factor value and calculated Darcy friction factor value are relatively close, so repeating the calculation using a friction factor of 0.018 will not change the pipe diameter to the next standard size (10 in, inside diameter of 9.976 in). Use 8 in diameter pipe.

The answer is (C).

Author Commentary

🕐 Stop performing iterative calculations when the calculated Darcy friction factor value is relatively close to the estimated value. It may seem more accurate to continue to calculate Darcy friction factor values until the calculated and estimated values are identical, but small changes in the Darcy friction factor value will not result in a change in the pipe size.

7. Determine the area of flow and hydraulic radius.

A	area of flow	ft^2
d_n	normal depth	ft
w	channel width	ft

$$A = d_n w = \left(\frac{8\ \text{in}}{12\ \dfrac{\text{in}}{\text{ft}}}\right)(5\ \text{ft}) = 3.33\ \text{ft}^2$$

R	hydraulic radius	ft

$$R = \frac{d_n w}{w + 2d_n} = \frac{\left(\dfrac{8\ \text{in}}{12\ \dfrac{\text{in}}{\text{ft}}}\right)(5\ \text{ft})}{5\ \text{ft} + (2)\left(\dfrac{8\ \text{in}}{12\ \dfrac{\text{in}}{\text{ft}}}\right)} = 0.526\ \text{ft}$$

The Manning roughness coefficient of concrete is 0.013.

n	Manning roughness coefficient	–
Q	flow rate	ft^3/sec
S	channel slope	%/100%

$$\begin{aligned}
Q &= \left(\frac{1.49}{n}\right) A R^{2/3} \sqrt{S} \\
&= \left(\frac{1.49}{0.013}\right)(3.33\ \text{ft}^2)(0.526\ \text{ft})^{2/3}\sqrt{\frac{0.2\%}{100\%}} \\
&= 11.12\ \text{ft}^3/\text{sec} \quad (11\ \text{ft}^3/\text{sec})
\end{aligned}$$

The answer is (B).

Author Commentary

🕐 The solution is a simple three-step calculation: area, hydraulic radius, and flow rate. For open channel flow, use the Manning equation for U.S. units.

💣 This problem involves three simple calculations, but pay careful attention to the slope term.

8. For 1:1 side slopes, the channel side slope angle measured from the horizontal is 45°.

d_1 upstream section water depth   m
R_1 upstream section hydraulic radius   m
θ channel side slope angle measured from the horizontal   deg

$$d_1 = \frac{10 \text{ cm}}{100 \; \frac{\text{cm}}{\text{m}}} = 0.1 \text{ m}$$

$$R_1 = \frac{d_1 \cos \theta}{2} = \frac{(0.1 \text{ m})(\cos 45°)}{2}$$
$$= 0.035 \text{ m}$$

The Manning roughness coefficient of concrete is 0.013.

n Manning roughness coefficient   –
S_1 upstream section channel slope   m/m
v_1 upstream section flow velocity   m/s

$$v_1 = \frac{R_1^{2/3} \sqrt{S_1}}{n}$$
$$= \frac{(0.035 \text{ m})^{2/3} \sqrt{0.12 \; \frac{\text{m}}{\text{m}}}}{0.013}$$
$$= 2.85 \text{ m/s}$$

Because equations for triangular channels are not standardized, the hydraulic jump equation must be derived from the momentum equation.

A area m^2
F force   N
g gravitational acceleration, 9.81 m/s^2
Q flow rate $\text{m}^3\text{/s}$
v_2 downstream section flow velocity   m/s
z depth to centroid   m
γ specific weight kN/m^3
ρ water density kg/m^3

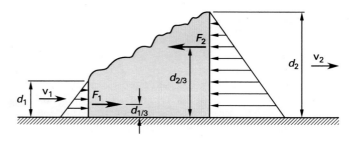

$$F_1 - F_2 = \rho Q v_2 - \rho Q v_1$$
$$\rho = \frac{\gamma}{g}$$
$$F = \gamma z A$$
$$Q = A v$$

For 1:1 triangular channels,

d water depth   m

$$A = d^2$$
$$z = \frac{d}{3}$$

Make substitutions into the momentum equation, and solve for the downstream section water depth.

d_2 downstream section water depth   m

$$Q = A_1 v_1 = d_1^2 v_1 = (0.1 \text{ m})^2 \left(2.85 \; \frac{\text{m}}{\text{s}} \right)$$
$$= 0.0285 \text{ m}^3\text{/s}$$
$$F_1 - F_2 = \rho Q v_2 - \rho Q v_1$$
$$\frac{d_1^3}{3} - \frac{d_2^3}{3} = \frac{Q^2}{g d_2^2} - \frac{Q^2}{g d_1^2}$$
$$\frac{(0.1 \text{ m})^3}{3} - \frac{d_2^3}{3} = \frac{\left(0.0285 \; \frac{\text{m}^3}{\text{s}} \right)^2}{\left(9.81 \; \frac{\text{m}}{\text{s}^2} \right) d_2^2} - \frac{\left(0.0285 \; \frac{\text{m}^3}{\text{s}} \right)^2}{\left(9.81 \; \frac{\text{m}}{\text{s}^2} \right)(0.1 \text{ m})^2}$$
$$d_2 = 0.28 \text{ m}$$

The answer is (B).

Author Commentary

🕐 Because this problem deals with a triangular channel, it may be difficult to find appropriate equations. Rather than spend time searching for the equations, it is quicker to derive the hydraulic jump equation from the momentum equation.

💣 Be aware that trapezoidal channel equations cannot be used when the channel is triangular.

9. For 1:1 side slopes, the side slope angle is 45°. For the channel,

A_1	channel wetted area	ft^2
b	channel base width	ft
d_1	channel water depth	ft
θ	side slope angle measured from the horizontal	deg

$$A_1 = \left(b + \frac{d_1}{\tan\theta}\right)d_1 = \left(8 \text{ ft} + \frac{3 \text{ ft}}{\tan 45°}\right)(3 \text{ ft})$$
$$= 33 \text{ ft}^2$$

R_1	channel hydraulic radius	ft

$$R_1 = \frac{A_1}{b + \frac{2d_1}{\sin\theta}} = \frac{33 \text{ ft}^2}{8 \text{ ft} + \frac{(2)(3 \text{ ft})}{\sin 45°}}$$
$$= 2.0 \text{ ft}$$

The Manning roughness coefficient of concrete is 0.013.

n	Manning roughness coefficient	–
Q	flow rate	ft^3/sec
S	slope	ft/ft
v_1	flow velocity	ft/sec

$$Q_1 = \left(\frac{1.49}{n}\right)A_1 R_1^{2/3}\sqrt{S}$$
$$= \left(\frac{1.49}{0.013}\right)(33 \text{ ft}^2)(2.0 \text{ ft})^{2/3}\sqrt{0.02 \, \frac{\text{ft}}{\text{ft}}}$$
$$= 849 \text{ ft}^3/\text{sec}$$

For the culvert,

A_2	culvert wetted area	ft^2
d_2	culvert water depth	ft
R_2	culvert hydraulic radius	ft
w	culvert width	ft

$$A_2 = wd_2 = (8 \text{ ft})d_2$$
$$R_2 = \frac{A_2}{w + 2d_2} = \frac{(8 \text{ ft})d_2}{8 \text{ ft} + 2d_2} = \frac{(4 \text{ ft})d_2}{4 + d_2}$$
$$Q_1 = Q_2 = \left(\frac{1.49}{n}\right)A_2 R_2^{2/3}\sqrt{S}$$
$$= 849 \text{ ft}^3/\text{sec}$$

Combine equations for velocity, flow, hydraulic radius, and area, and drop units for depth to simplify.

$$849 \, \frac{\text{ft}^3}{\text{sec}} = \left(\frac{1.49}{0.013}\right)8d_2\left(\frac{4d_2}{4 + d_2}\right)^{2/3}\sqrt{0.02}$$
$$6.5 = d_2\left(\frac{4d_2}{4 + d_2}\right)^{2/3}$$

Solve for the culvert water depth by trial and error.

$$d_2 = 4.1 \text{ ft}$$

The answer is (C).

Author Commentary

🕐 The depth calculation requires using either the solve function on a calculator or trial and error. Trial and error should produce a quick result if the channel depth of flow is used as the initial guess.

10. Find the flow velocity.

g	gravitational acceleration, 32.2	ft/sec^2
h	upstream head	ft
v	flow velocity	ft/sec

$$v = \sqrt{2gh} = \sqrt{(2)\left(32.2 \, \frac{\text{ft}}{\text{sec}^2}\right)(32 \text{ ft})}$$
$$= 45.4 \text{ ft/sec}$$

The criterion for baffled outlets usually sets the maximum flow velocity at 50 ft/sec. (See *Design of Small Canal Structures*, U.S. Department of the Interior.) Because 45.4 ft/sec is less than 50 ft/sec, a baffled outlet can be used.

Calculate the channel cross-sectional area.

A_{ch} discharge channel cross-sectional area ft^2

Q flow rate ft^3/sec

$$A_{ch} = \frac{Q}{v} = \frac{120 \ \frac{ft^3}{sec}}{45.4 \ \frac{ft}{sec}} = 2.643 \ ft^2$$

For a square channel,

D_{ch} flow depth in channel at outlet   ft

$$A_{ch} = D_{ch}^2$$
$$D_{ch} = \sqrt{A_{ch}} = \sqrt{2.643 \ ft^2} = 1.626 \ ft$$

Use the Froude number to relate the channel depth to the outlet basin depth.

D_b baffled outlet basin depth   ft
Fr   Froude number   –
w baffled outlet basin width   ft

$$Fr_1 = Fr_2$$
$$Fr = \frac{v}{\sqrt{gD_{ch}}} = \frac{Q}{D_b w \sqrt{gD_b}}$$
$$\frac{v}{\sqrt{gD_{ch}}} = \frac{Q}{D_b w \sqrt{gD_b}}$$
$$\frac{45.4 \ \frac{ft}{sec}}{\sqrt{\left(32.2 \ \frac{ft}{sec^2}\right)(1.626 \ ft)}} = \frac{120 \ \frac{ft^3}{sec}}{D_b(6 \ ft)\sqrt{\left(32.2 \ \frac{ft}{sec^2}\right)D_b}}$$
$$6.274 = 3.52 D_b^{3/2}$$
$$D_b = 0.68 \ ft$$

The answer is (B).

Author Commentary

🕐 This calculation may appear complex at first glance, but the key is to recognize how the Froude number ties the channel depth to the basin depth. Determine the channel depth, use that to calculate the Froude number, and use the Froude number to determine the basin depth.

💣 These calculations include exponent terms that are easily missed or misapplied.

11. Find the upstream flow rate.

d_1 upstream water depth   ft
Q flow rate ft^3/sec
v   flow velocity   ft/sec
w channel width   ft

$$Q = v d_1 w = \left(3 \ \frac{ft}{sec}\right)(8 \ ft)(15 \ ft)$$
$$= 360 \ ft^3/sec$$

Critical depth is the depth of flow that minimizes the energy flow.

d_c critical water depth over the barrier   ft
E specific energy   ft
g gravitational acceleration, 32.2 ft/sec^2
y_c critical barrier height   ft

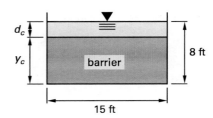

$$E = d_1 + \frac{Q^2}{2gw^2 d_1^2} = 8 \ ft + \frac{\left(360 \ \frac{ft^3}{sec}\right)^2}{(2)\left(32.2 \ \frac{ft}{sec^2}\right)(15 \ ft)^2 (8 \ ft)^2}$$
$$= 8.14 \ ft$$

$$d_c = \left(\frac{Q^2}{gw^2}\right)^{1/3} = \left(\frac{\left(360\ \dfrac{\mathrm{ft}^3}{\sec}\right)^2}{\left(32.2\ \dfrac{\mathrm{ft}}{\sec^2}\right)(15\ \mathrm{ft})^2}\right)^{1/3}$$

$$= 2.62\ \mathrm{ft}$$

$$d + \frac{Q^2}{2gw^2d^2} = d_c + \frac{Q^2}{2gw^2d_c^2} + y_c$$

$$8.14\ \mathrm{ft} = d_c + \frac{Q^2}{2gw^2d_c^2} + y_c$$

$$= 2.62\ \mathrm{ft} + \frac{\left(360\ \dfrac{\mathrm{ft}^3}{\sec}\right)^2}{(2)\left(32.2\ \dfrac{\mathrm{ft}}{\sec^2}\right)} + y_c$$
$$\times(15\ \mathrm{ft})^2(2.62\ \mathrm{ft})^2$$

$$y_c = 8.14\ \mathrm{ft} - 2.62\ \mathrm{ft} - 1.3\ \mathrm{ft}$$
$$= 4.2\ \mathrm{ft}$$

The answer is (D).

12. Find the depth of the water downstream of the hydraulic jump.

d_1	upstream water depth	m
d_2	downstream water depth	m
g	gravitational acceleration, 9.81	m/s²
v_1	upstream flow velocity	m/s

$$d_2 = -0.5d_1 + \sqrt{\frac{2v_1^2 d_1}{g} + \frac{d_1^2}{4}}$$

$$= -(0.5)(0.65\ \mathrm{m}) + \Bigg(\frac{(2)\left(8\ \dfrac{\mathrm{m}}{\mathrm{s}}\right)^2(0.65\ \mathrm{m})}{9.81\ \dfrac{\mathrm{m}}{\mathrm{s}^2}}$$
$$+ \frac{(0.65\ \mathrm{m})^2}{4}\Bigg)^{1/2}$$

$$= 2.6\ \mathrm{m}$$

q	flow rate per unit of channel width	m³/s·m
v_2	downstream flow velocity	m/s

$$q = v_1 d_1 = v_2 d_2 = \left(8\ \frac{\mathrm{m}}{\mathrm{s}}\right)(0.65\ \mathrm{m})$$

$$= 5.2\ \mathrm{m}^3/\mathrm{s{\cdot}m}\quad[\text{same as } \mathrm{m}^2/\mathrm{s}]$$

$$v_2 = \frac{q}{d_2} = \frac{5.2\ \dfrac{\mathrm{m}^3}{\mathrm{s{\cdot}m}}}{2.6\ \mathrm{m}}$$

$$= 2.0\ \mathrm{m/s}$$

ΔE	change in specific energy	m

$$\Delta E = d_1 + \frac{v_1^2}{2g} - d_2 - \frac{v_2^2}{2g}$$

$$= 0.65\ \mathrm{m} + \frac{\left(8\ \dfrac{\mathrm{m}}{\mathrm{s}}\right)^2}{(2)\left(9.81\ \dfrac{\mathrm{m}}{\mathrm{s}^2}\right)} - 2.6\ \mathrm{m}$$

$$- \frac{\left(2.0\ \dfrac{\mathrm{m}}{\mathrm{s}}\right)^2}{(2)\left(9.81\ \dfrac{\mathrm{m}}{\mathrm{s}^2}\right)}$$

$$= 1.11\ \mathrm{m}$$

The density of water is 1000 kg/m³.

q_m	mass flow rate per unit of channel width	kg/s·m
ρ	water density	kg/m³

$$q_m = q\rho = \left(5.2\ \frac{\mathrm{m}^3}{\mathrm{s{\cdot}m}}\right)\left(1000\ \frac{\mathrm{kg}}{\mathrm{m}^3}\right)$$

$$= 5200\ \mathrm{kg/s{\cdot}m}$$

P	power dissipated	kW/m

$$P = q_m g \Delta E$$
$$= \frac{\left(5200 \ \frac{\text{kg}}{\text{s·m}}\right)\left(9.81 \ \frac{\text{m}}{\text{s}^2}\right)(1.11 \ \text{m})}{1000 \ \frac{\text{kW}}{\text{W}}}$$
$$= 56.6 \ \text{kW/m} \quad (57 \ \text{kW/m})$$

The answer is (B).

Author Commentary

🕐 The power dissipation is related to the change in specific energy. Therefore, start with the energy equation. This will reveal which unknown velocity and depth values need to be calculated and help eliminate unnecessary calculations.

💣 The solution involves straightforward calculations, but requires the use of consistent units throughout. To ensure that units are consistent, show units in your calculations, and cancel common terms as you go.

13. From the problem illustration, the 60 min duration and 5.8 in/hr intensity storm coordinates fall on the 2% frequency curve. A 2% frequency storm is one that occurs two times in every 100 years, or once every 50 years.

The answer is (D).

Author Commentary

🕐 This problem does not require a calculation other than to convert the frequency as a percentage to an interval in years. The problem is solved by plotting the duration and intensity as shown in the illustration. Do not get bogged down by unnecessary calculations.

💣 Be careful when converting from a frequency given as a percentage to an interval given in years. It is a common mistake to assume that frequency as a percentage is the same as the interval in years. However, it is evident from the illustration that as the frequency increases, the intensity decreases. Therefore, a lower frequency represents a longer interval.

14. The annual probability of a flood event occurring is

n	period of interest	–
P_F	annual probability of a flood event	%
R	acceptable risk of a flood event occurring	–

$$R = \frac{1\%}{100\%} = 0.01$$
$$P_F = 1 - (1)(1 - R)^{1/n} = 1 - (1)(1 - 0.01)^{1/50}$$
$$= 0.00020 \quad (0.020\%)$$

The answer is (B).

Author Commentary

🕐 Though this is a simple solution, it is frequently done incorrectly by choosing the wrong probability equation.

💣 The question asks for the *annual* probability.

15. Each grid segment on the hydrograph represents a discharge volume of

$$\left(250,000 \ \frac{\text{ft}^3}{\text{sec}}\right)(2 \ \text{hr})\left(\frac{3600 \ \frac{\text{sec}}{\text{hr}}}{43,560 \ \frac{\text{ft}^2}{\text{ac}}}\right) = 41,322 \ \text{ac-ft}$$

duration interval (hr)	segments
4–8	8.5
8–12	37
12–16	32.5
16–20	20
20–24	11.5
24–28	5.5
28–32	1.5
	116.5

Q total discharge   in

$$Q = \frac{(116.5)(41{,}322 \text{ ac-ft})\left(12 \frac{\text{in}}{\text{ft}}\right)}{(5480 \text{ mi}^2)\left(640 \frac{\text{ac}}{\text{mi}^2}\right)}$$

$$= 16.47 \text{ in} \quad (16 \text{ in})$$

The answer is (D).

Author Commentary

🕐 Don't waste time by carefully separating base flow. Use a simple straight-line separation.

💣 There is no equation involved; the calculation is intuitive and requires manual integration. Count segments methodically to avoid confusion, and wait until the total segment count is determined before applying the volume calculation. This will keep the numbers small until the last step and make it easy to check the count.

16. Find where the observed fall and Q/Q_o intersect on the graph. For an observed fall of 0.3 m,

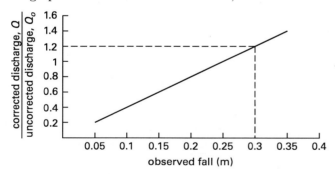

Q corrected discharge m^3/s
Q_o uncorrected discharge m^3/s

$$\frac{Q}{Q_o} = 1.2$$

From the illustration shown for a 2.3 m stage,

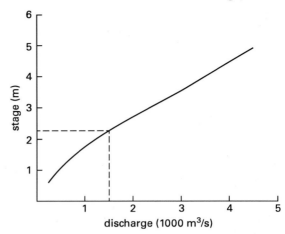

$$Q_o = 1500 \text{ m}^3/\text{s}$$

$$Q = \left(\frac{Q}{Q_o}\right)Q_o = (1.2)\left(1500 \frac{\text{m}^3}{\text{s}}\right)$$

$$= 1800 \text{ m}^3/\text{s}$$

The answer is (D).

Author Commentary

🕐 The observed fall curve shows that the corrected discharge will be greater than the discharge from the stage-discharge curve. The stage-discharge curve shows that the flow will be more than 1500 m^3/s. Therefore, before any calculations are done, it is evident from observation that the correct option is 1800 m^3/s, because it is the only option greater than 1500 m^3/s. However, it is good practice to do the calculations to check your work.

💣 The two curves must be used together to get the right answer. Be sure to include the observed fall correction.

17. Find the freezing elevation.

T_w temperature at the weather station   °C
z_f freezing point elevation   m
z_w weather station elevation   m
Γ ambient lapse rate   °C/m

$$z_f = z_w + \frac{0°\text{C} - T_w}{\Gamma} = 2036 \text{ m} + \frac{0°\text{C} - 12°\text{C}}{-0.016 \frac{°\text{C}}{\text{m}}}$$

$$= 2786 \text{ m}$$

For the problem illustration, the area between the snow line elevation, 1370 m, and the freezing point elevation, 2786 m, is the melting area.

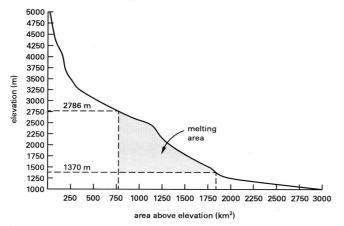

A_m melting area m^2

$$A_m = (1850 \text{ km}^2 - 755 \text{ km}^2)\left(1000 \ \frac{\text{m}}{\text{km}}\right)^2$$
$$= 1.1 \times 10^9 \text{ m}^2$$

T_s temperature at the snow line   °C
z_s snow line elevation   m

$$T_s = T_w + \Gamma(z_s - z_w)$$
$$= 12°\text{C} + \left(-0.016 \ \frac{°\text{C}}{\text{m}}\right)(1370 \text{ m} - 2036 \text{ m})$$
$$= 22.7°\text{C}$$

T_m average temperature over the melting   °C
           area

$$T_m = 0.5 \, T_s = (0.5)(22.7°\text{C}) = 11.3°\text{C}$$

d_m average degree days   °C·d

$$d_m = 11.3°\text{C} \cdot \text{d}$$

d_f degree-day factor   mm/°C·d
V_s snow melt volume m^3

$$V_s = d_m d_f A_m = \frac{(11.3°\text{C·d})\left(3 \ \dfrac{\text{mm}}{°\text{C} \cdot \text{d}}\right)(1.1 \times 10^9 \text{ m}^2)}{1000 \ \dfrac{\text{mm}}{\text{m}}}$$
$$= 3.73 \times 10^7 \text{ m}^3 \quad (3.7 \times 10^7 \text{ m}^3)$$

The answer is (B).

Author Commentary

🕐 The melting area is determined using the illustration shown in the problem statement.

💣 In the illustration shown in the problem statement, the snow line elevation is the lower elevation, but the upper elevation must be calculated. Don't make the mistake of using the station elevation. The upper elevation occurs some distance above the station as a function of the lapse rate.

18. The total runoff coefficient for the watershed is calculated using the area-weighted average.

A area   ac
C rational runoff coefficient   –

$$C_{\text{total}} = \frac{C_1 A_1 + C_2 A_2 + C_3 A_3}{A_1 + A_2 + A_3}$$
$$= \frac{(0.2)(12 \text{ ac}) + (0.6)(8 \text{ ac}) + (0.08)(17 \text{ ac})}{12 \text{ ac} + 8 \text{ ac} + 17 \text{ ac}}$$
$$= 0.23$$

Since the peak runoff coincides with the time to concentration, the rainfall intensity is 2.8 in/hr.

A_d drainage area   ac
I rainfall intensity   in/hr
Q_p peak runoff   ft³/sec

$$Q_p = C_{\text{total}} I A_d = (0.23)\left(2.8 \ \frac{\text{in}}{\text{hr}}\right)(12 \text{ ac} + 8 \text{ ac} + 17 \text{ ac})$$
$$= 23.8 \text{ ac-in/hr} \quad (24 \text{ ft}^3/\text{sec})$$

The answer is (B).

Author Commentary

🕐 Although the exact value for the conversion from ac-in/hr to ft^3/sec is 1.008, using the common approximation of 1.0 saves time.

💣 Because there are three adjacent tracts, each with a different runoff coefficient, a weighted coefficient calculation is required. Using a simple average of the runoff coefficients will overestimate the peak flow.

19. Find the difference between the observation well drawdown and the aquifer thickness.

h	h_o – observation well drawdown	ft
h_o	aquifer thickness	ft

$$h = 18 \text{ ft} - \frac{14 \text{ in}}{12 \frac{\text{in}}{\text{ft}}}$$
$$= 16.83 \text{ ft}$$

K	hydraulic conductivity	ft/day
Q	pumping rate	gpm
r	distance from pumped well to observation well	ft
r_o	radius of influence	ft

$$Q = \frac{K\pi(h_o^2 - h^2)}{\ln\frac{r_o}{r}}$$

$$20 \frac{\text{gal}}{\text{min}} = \frac{\left(7.2 \frac{\text{ft}}{\text{day}}\right)\pi((18 \text{ ft})^2 - (16.83 \text{ ft})^2)}{\ln\left(\frac{r_o}{30 \text{ ft}}\right)\left(1440 \frac{\text{min}}{\text{day}}\right)}$$

$$r_o = 38 \text{ ft}$$

The answer is (C).

Author Commentary

🕐 The problem involves an unconfined aquifer. Under steady-state conditions, the radius of influence is not a function of time. This information indicates that the steady-state drawdown equation for a fully penetrating well in an unconfined aquifer should be used. Recognizing these hints in problem statements will save time.

💣 The only places in this solution where errors are likely are in unit conversions and in calculating using the natural log. Double check these.

20. Using the soil classification, determine the minimum typical hydraulic conductivity for each layer.

d_i	thickness of layer i	cm
K	hydraulic conductivity	cm/s

layer	thickness, d (cm)	soil class	K (cm/s)	d/K (s)
1	70	SP	5×10^{-4}	1.40×10^5
2	109	GC	5×10^{-8}	2.18×10^9
3	88	SM	2.5×10^{-5}	3.52×10^6
4	46	SC	2.5×10^{-7}	1.84×10^8

$$K_{\text{overall}} = \frac{d_1 + d_2 + \cdots + d_n}{\frac{d_1}{K_1} + \frac{d_2}{K_2} + \cdots + \frac{d_n}{K_n}}$$
$$= \frac{70 \text{ cm} + 109 \text{ cm} + 88 \text{ cm} + 46 \text{ cm}}{(1.40 \times 10^5 \text{ s}) + (2.18 \times 10^9 \text{ s}) + (3.52 \times 10^6 \text{ s}) + (1.84 \times 10^8 \text{ s})}$$
$$= 1.3 \times 10^{-7} \text{ cm/s}$$

The answer is (B).

Author Commentary

🕐 Only the soil classification is given. Try using typical hydraulic conductivity values to estimate the conductivity of each layer.

💣 Some terms in the solution are represented by large numbers or exponents. Adding and multiplying these is a common source of error.

21. Dispersion and diffusion cause dilution of the solute. In more permeable materials, dispersion is far more significant. Diffusion becomes more important in less permeable soils and may occur even where groundwater flow is zero.

Dispersion is caused by mixing in the pore spaces and by differential flow in flow channels, a phenomenon that becomes more significant as the gradient and permeability of the soil increase. Where dispersion dominates, the result is a more pronounced elongated plume that moves away from the source down the gradient.

Diffusion accounts for the molecular spreading of the solute across the boundary between relatively contaminated water and clean water. The fluid does not have to be moving for diffusion to occur. In more permeable media, diffusion is most pronounced in its impact at the boundary. It tends to flatten the concentration gradient across the boundary. In low permeability soils, however, diffusion may create a radial plume around the source.

For the conditions described in the problem statement, diffusion will likely dominate, and the plume will be relatively circular in shape.

The answer is (D).

Author Commentary

🕐 Read the problem statement carefully to fully understand the described conditions. The solution is in the definitions of dispersion and diffusion. Find these definitions and apply them to the described conditions.

💣 Diffusion and dispersion are frequently confused. Be sure to apply the correct definition to each.

22. The average depth of flow is a function of the average flow rate.

d average water depth between   in
        manholes

$$d = \frac{3.8 \text{ in} + 5.2 \text{ in}}{2}$$
$$= 4.5 \text{ in}$$

Pipe diameters are inside diameters.

D pipe diameter   in
d/D partial-to-full flow ratio   —

$$\frac{d}{D} = \frac{4.5 \text{ in}}{16 \text{ in}}$$
$$= 0.28$$

From standard tables for areas of partially full pipes, based on the partial-to-full flow ratio,

A average cross-sectional area of flow ft^2

$$A = 0.32 \text{ ft}^2$$

t average travel time between manholes   sec

$$t = \frac{398 \text{ sec} + 512 \text{ sec}}{2} = 455 \text{ sec}$$

Q flow rate ft^3/sec
x distance between manholes   ft

$$Q = \frac{Ax}{t}$$
$$= \frac{(0.32 \text{ ft}^2)(1076 \text{ ft})}{455 \text{ sec}}$$
$$= 0.76 \text{ ft}^3/\text{sec}$$

The answer is (B).

Author Commentary

🕐 The problem statement gives upstream and downstream values for depth of flow. Although it is possible to get a solution by doing two sets of calculations and averaging the flow rates, a quicker solution is to average the depths of flow and then perform a single flow calculation. Save time by using tables to determine areas for partially full pipes.

💣 The calculations are simple addition, subtraction, multiplication, and division functions. Errors will likely occur if inconsistent units are used. Avoid this by converting all distance units as a first step.

23. The BOD_5 of the effluent suspended solids is

f biodegradable fraction of total suspended solids (TSS)   –

G stoichiometric oxygen demand for cell oxidation   g/g

S_o influent BOD_5 mg/L

S_u influent BOD_u mg/L

S_x BOD_5 of effluent suspended solids   mg/L

X_e effluent TSS   mg/L

$$S_x = fX_e G\frac{S_o}{S_u} = (0.62)\left(18\ \frac{mg}{L}\right)\left(1.42\ \frac{g}{g}\right)\left(\frac{217\ \frac{mg}{L}}{312\ \frac{mg}{L}}\right)$$
$$= 11\ mg/L$$

S soluble influent BOD_5 escaping treatment   mg/L

S_e effluent BOD_5 mg/L

$$S = S_e - S_x = 20\ \frac{mg}{L} - 11\ \frac{mg}{L} = 9.0\ mg/L$$

The answer is (B).

Author Commentary

⏰ Determine the relationship between effluent solids and effluent BOD. It may be necessary to develop a formula from basic principles. Include the biodegradable fraction of total suspended solids and the stoichiometric oxygen demand for cell oxidation. Do this calculation first.

💣 The temptation is to simply conclude that the effluent BOD is BOD that has escaped treatment. However, the problem asks for *soluble* BOD. Total BOD includes solids that also exert BOD, while soluble BOD does not include these solids. Using the total BOD that has escaped treatment results in an overestimation of the soluble BOD concentration.

24. The wasted sludge mass is the product of the wasted solids flow rate and the wasted solids concentration, $Q_w X_u$, which is equal to 34 kg/d. The bioreactor sludge mass is

Q_w wasted solids flow rate m^3/d

SRT_c solids residence time   d

V reactor volume m^3

X biomass concentration in the reactor   mg/L

X_u wasted solids concentration   mg/L

$$SRT_c = \frac{VX}{Q_w X_u}$$
$$VX = (SRT_c)(Q_w X_u)$$
$$= (10\ d)\left(34\ \frac{kg}{d}\right)$$
$$= 340\ kg$$

F/M food-to-microorganism ratio d^{-1}

Q influent wastewater flow rate   L/d

S_o influent wastewater BOD   kg/L

$$\frac{F}{M} = \frac{QS_o}{VX}$$
$$= \frac{\left(18\,925\ \frac{m^3}{d}\right)\left(247\ \frac{mg}{L}\right)\times\left(1000\ \frac{L}{m^3}\right)\left(10^{-6}\ \frac{kg}{mg}\right)}{340\ kg}$$
$$= 13.7\ d^{-1}\quad(14\ d^{-1})$$

The answer is (B).

Author Commentary

⏰ The tricky part of this solution is finding the bioreactor sludge mass. Bioreactor sludge mass is part of the definition of solids residence time (SRT) and can be found from the SRT given. Knowing the basic definition of SRT can save time and avoid lengthy calculations and searches for information.

💣 Don't confuse the wasted sludge mass and the bioreactor sludge mass.

25. For nitrification,

k_d	endogenous decay coefficient	d^{-1}
K_s	half velocity constant	mg/L
N_o	influent ammonia nitrogen concentration	mg/L
SRT_m	minimum solids residence time	d
μ_m'	maximum specific bacterial growth rate for nitrification	d^{-1}

$$\frac{1}{\text{SRT}_m} = \frac{\mu_m' N_o}{K_s + N_o} - k_d$$
$$= \frac{\left(\dfrac{0.41}{\text{d}}\right)\left(51\ \dfrac{\text{mg}}{\text{L}}\right)}{2.6\ \dfrac{\text{mg}}{\text{L}} + 51\ \dfrac{\text{mg}}{\text{L}}} - \frac{0.07}{\text{d}}$$
$$= 0.32\ \text{d}^{-1}$$
$$\text{SRT}_m = 3.1\ \text{d}$$

For BOD removal,

S_o	influent BOD concentration	mg/L
μ_m	maximum growth rate constant	d^{-1}

$$\frac{1}{\text{SRT}_m} = \frac{\mu_m S_o}{K_s + S_o} - k_d = \frac{\left(\dfrac{0.50}{\text{d}}\right)\left(312\ \dfrac{\text{mg}}{\text{L}}\right)}{2.6\ \dfrac{\text{mg}}{\text{L}} + 312\ \dfrac{\text{mg}}{\text{L}}} - \frac{0.07}{\text{d}}$$
$$= 0.43\ \text{d}^{-1}$$
$$\text{SRT}_m = 2.3\ \text{d}$$

The greatest minimum solids residence controls design. Nitrification controls design with a solids residence time of 3.1 d.

The answer is (C).

Author Commentary

🕐 The solution involves two similar calculations. Recognizing and exploiting the similarities in problems like these will save time.

💣⁜ Solids residence times (SRT) can be calculated for both BOD and nitrogen removal. Whichever is greatest will control design. Be sure to perform both calculations.

26. The Clean Water Act (CWA) primarily addresses discharges to waters of the United States by imposing effluent limitations through a pretreatment and permit program. Three major areas of the CWA are discharge criteria, permitting, and priority pollutants.

Discharge criteria provisions set pretreatment standards for categorical discharges from industry to publicly owned treatment works (POTWs). Discharger categories are specifically defined in the regulations and apply to nondomestic waste generators who discharge to a POTW. Discharge criteria also set secondary treatment standards applicable to discharges to receiving waters from POTWs. These include treatment standards for specific "conventional pollutants" such as biochemical oxygen demand (BOD), suspended solids (SS), and pH.

Permitting provisions define the National Pollutant Discharge Elimination System (NPDES), which regulates discharges to waters of the United States. In most cases, the NPDES program is administered by individual states. The NPDES permits consider site-specific conditions in establishing discharge criteria.

The 1977 amendments to the CWA included a list of 65 priority pollutants (specific chemicals and classes of chemicals) to be used for defining toxic substances and establishing permit limits. The original list has been expanded to the current list of 126 priority pollutants. Priority pollutants are those chemicals with relatively high toxicity and high production volume.

The Safe Drinking Water Act (SDWA) defines, among other things, maximum contaminant levels (MCLs) for drinking water. The MCLs are frequently used to establish clean-up levels in groundwater contamination remediation programs. The MCLs are not part of the CWA.

The answer is (B).

Author Commentary

💣⁜ Be careful not to confuse the Clean Water Act (CWA) with the Safe Drinking Water Act (SDWA).

27. Convert the flow rate to ft^3/min.

Q_d	flow rate	gal/day or ft^3/min

$$Q_d = \frac{0.86 \times 10^6\ \dfrac{\text{gal}}{\text{day}}}{\left(1440\ \dfrac{\text{min}}{\text{day}}\right)\left(7.48\ \dfrac{\text{gal}}{\text{ft}^3}\right)} = 80\ \text{ft}^3/\text{min}$$

Find the surface area required for thickening.

A_t	surface area required for thickening	ft^2
h	settling column height	ft
t_s	time to thickening	min

$$A_t = \frac{Q_d t_s}{h} = \frac{\left(80 \ \frac{ft^3}{min}\right)(21 \ min)}{6 \ ft}$$
$$= 280 \ ft^2$$

Find the surface area required for clarification.

h_u	settled solids height in column at t_s	ft
X	initial solids concentration	mg/L
X_u	thickened solids concentration	mg/L

$$h_u = \frac{Xh}{X_u} = \frac{\left(2600 \ \frac{mg}{L}\right)(6 \ ft)}{10{,}000 \ \frac{mg}{L}}$$
$$= 1.56 \ ft$$

Q_c	clarification flow rate	ft^3/min

$$Q_c = \frac{Q_d(h - h_u)}{h}$$
$$= \frac{\left(80 \ \frac{ft^3}{min}\right)(6 \ ft - 1.56 \ ft)}{6 \ ft}$$
$$= 59.2 \ ft^3/min$$

The slope of the settling curve is equal to the settling velocity.

A_c	surface area for clarification	ft^2
v_s	settling velocity	ft/min

$$A_c = \frac{Q_c}{v_s} = \frac{59.2 \ \frac{ft^3}{min}}{0.23 \ \frac{ft}{min}}$$
$$= 257 \ ft^2 \quad [< 280 \ ft^2]$$

The surface area required for thickening controls.

The answer is (C).

Author Commentary

🕐 This problem requires interpretation of the test results. Separate these results into their two parts, clarification and thickening, to clearly determine which information is relevant to each calculation. This will organize the test results and allow for more efficient calculations.

💣 The area is calculated based on clarification and thickening. Whichever is greatest will control design. Therefore, be sure to perform both calculations.

28. At the discharge point, assume influences on dissolved oxygen (DO) from reaeration and microbial activity are negligible.

Draw a mass balance diagram showing the inflow and outflow rates and concentrations.

DO_d	downstream dissolved oxygen concentration	mg/L
DO_u	upstream dissolved oxygen concentration	mg/L
DO_W	wastewater treatment plant dissolved oxygen concentration	mg/L
Q_d	downstream flow	ft^3/sec or L/s
Q_u	upstream flow	ft^3/sec or L/s
Q_W	wastewater treatment plant flow	ft^3/sec or L/s

$$Q_d = Q_u + Q_W$$
$$= \left(2000 \ \frac{ft^3}{sec}\right)\left(28.25 \ \frac{L}{ft^3}\right)$$
$$+ \frac{\left(15 \times 10^6 \ \frac{gal}{day}\right)\left(3.785 \ \frac{L}{gal}\right)}{86{,}400 \ \frac{sec}{day}}$$
$$= 57\,157 \ L/s$$

$\dot{m}_d$ downstream mass flow rate of dissolved oxygen   mg/s

$\dot{m}_u$ upstream mass flow rate of dissolved oxygen   mg/s

$\dot{m}_W$ wastewater treatment plant mass flow rate of dissolved oxygen   mg/s

$$\dot{m}_u = Q_u \text{DO}_u$$
$$= \left(2000 \ \frac{\text{ft}^3}{\text{sec}}\right)\left(8.10 \ \frac{\text{mg}}{\text{L}}\right)\left(28.25 \ \frac{\text{L}}{\text{ft}^3}\right)$$
$$= 457\,650 \ \text{mg/s}$$

$$\dot{m}_W = Q_W \text{DO}_W$$
$$= \frac{\left(15 \times 10^6 \ \frac{\text{gal}}{\text{day}}\right)\left(1.20 \ \frac{\text{mg}}{\text{L}}\right)\left(3.785 \ \frac{\text{L}}{\text{gal}}\right)}{86{,}400 \ \frac{\text{sec}}{\text{day}}}$$
$$= 789 \ \text{mg/s}$$

$$\dot{m}_d = \dot{m}_u + \dot{m}_W = 457\,650 \ \frac{\text{mg}}{\text{s}} + 789 \ \frac{\text{mg}}{\text{s}}$$
$$= 458\,439 \ \text{mg/s}$$

$$\text{DO}_d = \frac{\dot{m}_d}{Q_d} = \frac{458\,439 \ \frac{\text{mg}}{\text{s}}}{57\,157 \ \frac{\text{L}}{\text{s}}}$$
$$= 8.02 \ \text{mg/L}$$

The answer is (B).

Author Commentary

🕐   Look at this as a mass balance problem, and draw a sketch showing inputs and outputs. This is a quick way to understand the problem and will make it possible to rapidly develop the mass balance equations needed for the solution.

💣   Mass must be conserved throughout the solution. Ensuring that this is the case will help identify and correct any errors.

29. The 1977 amendments to the Clean Water Act (CWA) included a list of 65 priority pollutants (specific chemicals and classes of chemicals) to be used for defining toxic substances. The original list has been expanded to the current list of 126 priority pollutants. As defined by the CWA, priority pollutants are those chemicals with relatively high toxicity and high production volume. Although they may not be the most toxic chemicals, because of their generally widespread use and

relative toxicity they have received special attention from regulators.

The answer is (B).

Author Commentary

💣   Do not confuse the provisions of the Clean Water Act (CWA) with those of other similar regulations, such as the Resource Conservation and Recovery Act (RCRA) or the Comprehensive Environmental Response, Compensation, and Liability Act (CERCLA).

30. Assimilative capacity is the ability of the environment to absorb waste discharges and is influenced by both stock and fund pollutants. Stock pollutants are substances or materials for which the assimilative capacity is very small—essentially any discharge results in an unacceptable negative impact. Examples of stock pollutants are dioxins and lead, which are compounds that are toxic and that accumulate with little or very slow degradation. Fund pollutants are substances or materials for which the assimilative capacity is relatively large. These are compounds that degrade to produce little accumulation in the environment over time. Examples of fund pollutants are human and animal waste, typically measured by biochemical oxygen demand (BOD) and volatile suspended solids (VSS).

The answer is (D).

Author Commentary

💣   A correct answer depends on understanding the differences between stock and fund pollutants. Consult references if these differences are not clear.

31. Calculate and then graph the 96 h mortality rate at each concentration.

concentration (%)	survivors at 96 h	mortality at 96 h (%)
2	19	5
4	12	40
8	7	65
16	0	100

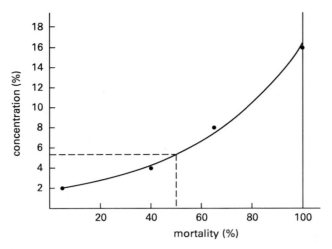

From the illustration, the 96 h LC50 occurs at 5.4% of predilution concentration.

$$96 \text{ h LC50} = (\text{predilution concentration})$$

$$\times \left(\frac{\begin{array}{c} \% \text{ of predilution concentration} \\ \text{at 50\% mortality} \end{array}}{100\%} \right)$$

$$= \left(1.63 \ \frac{\text{mg}}{\text{L}} \right) \left(\frac{5.4\%}{100\%} \right)$$

$$= 0.088 \text{ mg/L}$$

The answer is (A).

Author Commentary

🕐 Plot the test results using concentration and mortality as the axes of a graph. To save time, only plot the data for values near 50% mortality—data for concentrations of 4% and 8%.

32. The Thomas equation provides an adequate estimate of the most probable number (MPN).

$$\frac{\text{MPN}}{100 \text{ mL}} = \frac{(100)(\text{positives})}{\sqrt{\text{mL negatives}} \sqrt{\begin{array}{c} \text{mL positives} \\ + \text{mL negatives} \end{array}}}$$

$$= \frac{(100)(3 + 2 + 1)}{\sqrt{\begin{array}{c}(2)(1 \text{ mL}) + (3)(0.1 \text{ mL}) \\ +(4)(0.01 \text{ mL})\end{array}}}$$

$$\times \sqrt{\begin{array}{c}(5)(1 \text{ mL}) + (5)(0.1 \text{ mL}) \\ +(5)(0.01 \text{ mL})\end{array}}$$

$$= 166/100 \text{ mL} \quad (170/100 \text{ mL})$$

The answer is (D).

Author Commentary

🕐 Use a simple calculation to approximate MPN. The Thomas equation is a reliable, relatively straightforward method of approximating MPN.

33. Find the equivalent concentrations of the ions, and sum the concentrations of the cations and anions.

ion	concentration, C (mg/L)	valence, v (eq/mol)	mole weight, MW (g/mol)	MW/v (mg/meq)	equivalent concentration, C_{eq} (meq/L)
Ca^{++}	128	2	40	20	6.4
Mg^{++}	66	2	24	12	5.5
SO_4^{2-}	83	2	96	48	1.7
Cl^-	21	1	35	35	0.60
NO_3^-	14	1	62	62	0.23
HCO_3^-	279	1	61	61	4.6
Na^+	7	1	23	23	0.30

$$\sum \text{cations} = 6.4 \ \frac{\text{meq}}{\text{L}} + 5.5 \ \frac{\text{meq}}{\text{L}} + 0.30 \ \frac{\text{meq}}{\text{L}}$$

$$= 12.2 \text{ meq/L}$$

$$\sum \text{anions} = 1.7 \ \frac{\text{meq}}{\text{L}} + 0.60 \ \frac{\text{meq}}{\text{L}}$$

$$+ 0.23 \ \frac{\text{meq}}{\text{L}} + 4.6 \ \frac{\text{meq}}{\text{L}}$$

$$= 7.13 \text{ meq/L}$$

$$\sum \text{cations} \gg \sum \text{anions}$$

Not all ions that are likely present at a significant concentration in the sample are included in the analysis. It is deficient in anions.

The answer is (C).

Author Commentary

🕐 Concentrations should be converted to units of milliequivalents per liter. The calculations for converting milligrams per liter to milliequivalents per liter are repetitive. Save time by extending the table in the problem statement and working the calculations for each ion in sequence.

💣 No conclusions about the adequacy of the analysis can be made by adding the concentration of anions and cations in milligrams per liter.

34. For design flow based on population,

P population   people

$$P = (346 \text{ houses})\left(4 \ \frac{\text{people}}{\text{house}}\right) = 1384 \text{ people}$$

The average annual per capita daily flow is 165 gal/person-day.

q per capita flow demand   gal/person-day
Q_a average domestic flow demand   gpm

$$Q_a = Pq$$
$$= (1384 \text{ people})\left(\frac{165 \ \dfrac{\text{gal}}{\text{person-day}}}{1440 \ \dfrac{\text{min}}{\text{day}}}\right)$$
$$= 159 \text{ gpm}$$

For design flow for domestic uses during fire demand, the maximum daily flow multiplier is 1.5.

M peak daily flow multiplier   –
Q_p peak daily domestic flow demand   gpm

$$Q_p = Q_a M = \left(159 \ \frac{\text{gal}}{\text{min}}\right)(1.5)$$
$$= \left(159 \ \frac{\text{gal}}{\text{min}}\right)(1.5)$$
$$= 239 \text{ gpm}$$

The highest fire demand is the fire demand based on dwelling type, 1250 gpm. Use the fire demand based on dwelling type combined with peak daily flow to calculate the design flow.

Q_d design flow   gpm
Q_{fA} fire demand based on dwelling type   gpm

$$Q_d = Q_{fA} + Q_p$$
$$= 1250 \ \frac{\text{gal}}{\text{min}} + 239 \ \frac{\text{gal}}{\text{min}}$$
$$= 1489 \text{ gpm} \quad (1500 \text{ gpm})$$

The answer is (D).

Author Commentary

🕐 Three fire demand criteria are provided. Ignore the two lower demands and base design on the highest fire demand, in this case the demand based on dwelling type.

💣 Apply the peak multiplier to the domestic flow before including fire demand. Applying the peak multiplier to the fire demand will result in an overestimate of the design flow.

35. The theoretical hydraulic detention time is

Q flow rate m^3/d
t theoretical hydraulic detention time   h
V volume m^3

$$t = \frac{V}{Q} = \frac{(2.5 \text{ m})(15 \text{ m})(3.0 \text{ m})\left(24 \ \dfrac{\text{h}}{\text{d}}\right)}{900 \ \dfrac{m^3}{d}} = 3.0 \text{ h}$$

The actual hydraulic detention time is

E fractional efficiency   –
t_a actual hydraulic detention time   h

$$t_a = tE = (3.0 \text{ h})\left(\frac{83\%}{100\%}\right) = 2.5 \text{ h}$$

The answer is (B).

Author Commentary

💣※ The theoretical and actual hydraulic detention times are not the same. Applying an efficiency correction to the theoretical value gives the actual detention time, which is less than the theoretical detention time.

36. Find the hydraulic residence time.

G	average velocity gradient	s^{-1}
G_t	average time-velocity gradient	–
t	hydraulic residence time	s

$$t = \frac{G_t}{G} = \frac{4.5 \times 10^4}{\dfrac{40}{s}} = 1125 \text{ s}$$

| Q | flow rate | m^3/d |
| V | mixing basin volume | m^3 |

$$V = Qt = \frac{\left(18\,000\ \dfrac{m^3}{d}\right)(1125\text{ s})}{86\,400\ \dfrac{s}{d}}$$
$$= 234 \text{ m}^3$$

For horizontally mounted paddles, the most efficient section for the paddles to rotate within will be a square. The length of each section will equal the depth.

d	basin depth	m
L	basin total length	m
L_s	section length	m

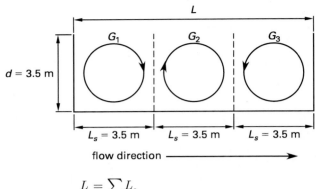

$$L = \sum L_s$$
$$= 3.5 \text{ m} + 3.5 \text{ m} + 3.5 \text{ m}$$
$$= 10.5 \text{ m}$$

| w | basin width | m |

$$w = \frac{V}{Ld} = \frac{234 \text{ m}^3}{(10.5 \text{ m})(3.5 \text{ m})} = 6.4 \text{ m} \quad (6.5 \text{ m})$$

The answer is (A).

Author Commentary

🕐 A labeled sketch of the longitudinal cross section of the basin may help with the visualization of the depth of the tank, the number of sections, and the orientation of the paddles.

💣※ In practice, the velocity gradient will vary from section to section throughout the basin, but an average gradient can be used in this calculation. This will reduce the number of calculation steps and chances for error without affecting the solution.

37. The flow rate can be found from the stripping factor equation using the dimensionless Henry's constant.

H_D	dimensionless Henry's constant	–
MW	molecular weight	g/mol
R^*	universal gas constant, 8.250×10^{-5}	atm·m³/ mol·K
S_w	solubility in water	mg/L
T	temperature	K
V_p	vapor pressure	atm

$$H_D = \frac{V_p(\text{MW})}{R^* T S_w}$$

$$= \frac{(0.26 \text{ atm})\left(112 \dfrac{\text{g}}{\text{mol}}\right)\left(1000 \dfrac{\text{mg}}{\text{g}}\right)\left(0.001 \dfrac{\text{m}^3}{\text{L}}\right)}{\left(8.250 \times 10^{-5} \dfrac{\text{atm·m}^3}{\text{mol·K}}\right)(20°\text{C} + 273°)\left(7300 \dfrac{\text{mg}}{\text{L}}\right)}$$

$$= 0.166$$

The required air flow rate is

G	gas loading rate	L/s
L	liquid loading rate	L/s
R	stripping factor	–

$$R = \frac{H_D G}{L}$$

$$3.5 = \frac{0.166 G}{37 \dfrac{\text{L}}{\text{s}}}$$

$$G = 780 \text{ L/s}$$

The answer is (C).

Author Commentary

🕐 The dimensionless Henry's constant can be calculated in two steps, or in a single equation that combines the two steps. Save time by using the single-step calculation. Additional time can be saved by choosing a universal gas law constant with units that are already compatible with the solubility and vapor pressure units.

💣 The solution requires a dimensionless Henry's constant. Using any other Henry's constant will give an incorrect result with irreconcilable units.

38. Find the equivalent concentration of Zn^{++} in the water. Zinc has a molecular weight of 65 g/mol and an equivalence of 2 eq/mol.

C	concentration	mg/L or eq/L
E	equivalence	eq/mol
MW	molecular weight	mg/mol

$$C_{\text{eq/L}} = \frac{C_{\text{mg/L}} E}{\text{MW}} = \frac{\left(27 \dfrac{\text{mg}}{\text{L}}\right)\left(2 \dfrac{\text{eq}}{\text{mol}}\right)}{\left(65 \dfrac{\text{g}}{\text{mol}}\right)\left(1000 \dfrac{\text{mg}}{\text{g}}\right)}$$

$$= 0.000831 \text{ eq/L}$$

Q	flow rate	m³/d
R	resin capacity	eq/L
$V_{\text{exchange resin}}$	daily resin volume	m³/d

$$V_{\text{exchange resin}} = \frac{Q C_{\text{eq/L}}}{R}$$

$$= \frac{\left(675 \dfrac{\text{m}^3}{\text{d}}\right)\left(0.000831 \dfrac{\text{eq}}{\text{L}}\right)}{0.93 \dfrac{\text{eq}}{\text{L}}}$$

$$= 0.603 \text{ m}^3/\text{d} \quad (0.60 \text{ m}^3/\text{d})$$

The answer is (B).

Author Commentary

🕐 The resin capacity is expressed in equivalents per liter. Converting the concentration of Zn^{++} to equivalents per liter before calculating the resin volume will save time.

💣 The solution requires converting from mass to equivalent concentration. The concentration and resin capacity units should cancel.

39. Using interpolation, determine the incremental removal efficiency for each corresponding depth for the selected time-depth pair.

t_o	settling time	min
Z_i	incremental depth	m
Z_o	design depth	m
$\Delta\eta_i$	incremental removal efficiency	%
η_o	minimum removal efficiency at design time-depth	%

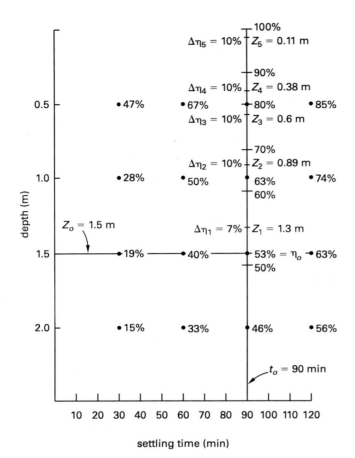

settling time (min)

At a settling time of 90 min and a settling zone depth of 1.5 m, the minimum removal efficiency is 53%, and the data can be tabulated as shown.

i	Z_i	$\Delta\eta_i$	$Z_i\Delta\eta_i$
1	1.3 m	0.07	0.091 m
2	0.89 m	0.10	0.089 m
3	0.60 m	0.10	0.060 m
4	0.38 m	0.10	0.038 m
5	0.11 m	0.10	0.011 m
		$\sum Z_i\Delta\eta_i$	= 0.289 m

η overall removal efficiency   %

$$\eta = \eta_o + \left(\frac{\sum Z_i\Delta\eta_i}{Z_o}\right)_{\%}$$

$$= 53\,\% + \left(\frac{0.289\text{ m}}{1.5\text{ m}} \times 100\%\right)$$

$$= 72.3\% \quad (72\%)$$

The answer is (C).

Author Commentary

🕐 Focus on the results corresponding to the 90 min settling time. Do not be distracted by the rest of the illustration.

💣※ From inspection of the illustration, it is apparent that the overall efficiencies will be greater than 53% but less than 97%. Therefore, ignore answer options A and D. Solutions close to these values indicate an error was made.

40. Use a waste sample of 100 lbm so that discarded mass can be expressed in pounds mass.

The dry mass is

$$m_{\text{dry}} = m_{\text{discarded}}\left(\frac{100\% - \text{discarded moisture in }\%}{100\%}\right)$$

The volume at 40% moisture is

$$V_{40\%,\text{moist}} = \frac{m_{\text{dry}}}{\rho_{\text{dry,discarded}}(1 - 0.4)}$$

Tabulate the dry mass and volume at 40% moisture.

component	discarded mass, $m_{\text{discarded}}$ (lbm)	discarded moisture (%)	dry mass, m_{dry} (lbm)	discarded dry density, $\rho_{\text{dry,discarded}}$ (lbm/ft^3)	volume at 40% moisture, $V_{40\%,\text{moist}}$ (ft^3)
paper	38	6	36	5	12
garden	25	60	10	7	2.4
food	9	70	2.7	18	0.25

The volume that paper should contribute to the mulch at 10% (by volume) is

$$V_{\text{paper}} = (2.4\text{ ft}^3 + 0.25\text{ ft}^3)(0.1) = 0.27\text{ ft}^3 < 12\text{ ft}^3$$

Paper is not limiting.

The mulch volume per 100 lbm waste is

$$V_{\text{mulch}} = 2.4\text{ ft}^3 + 0.25\text{ ft}^3 + 0.27\text{ ft}^3 = 2.9\text{ ft}^3$$

The potential revenue is

$$
\text{revenue} = (100{,}000 \text{ people})\left(5 \ \frac{\text{lbm waste}}{\text{person-day}}\right)
$$

$$
\times \left(\frac{\dfrac{2.9 \text{ ft}^3 \text{ mulch}}{100 \text{ lbm waste}}}{27 \ \dfrac{\text{ft}^3}{\text{yd}^3}}\right)\left(365 \ \frac{\text{day}}{\text{yr}}\right)\left(\frac{\$10}{1 \text{ yd}^3}\right)
$$

$$
= \$1.96 \times 10^6/\text{yr} \quad (\$2.0 \text{ million/yr})
$$

The answer is (A).

Author Commentary

🕐 Find the volume at 40% moisture for the separate mulch components, and sum them to determine the total mulch volume. The fastest way to do this is by extending the table to include columns for dry mass and volume at 40% moisture. Once the table is complete, the mulch volume and potential revenue are easily calculated.

💣 The units will guide the potential revenue calculation. If the units work, then the calculation is likely to be correct. The most frequent errors are made in the calculations for the volumes of the components. Be careful applying 40% moisture to the calculations; think of 40% moisture as equivalent to 60% dry.

Solutions
Practice Exam 2

41. Find the surface area of the gate.

A	gate surface area	m^2
D	gate diameter	m

$$A = \frac{\pi D^2}{4} = \frac{\pi (0.5 \text{ m})^2}{4} = 0.20 \text{ m}^2$$

| h_2 | water depth above the gate bottom | m |

$$h_2 = 2.3 \text{ m} + 0.5 \text{ m} = 2.8 \text{ m}$$

The density of water is 1000 kg/m³.

g	gravitational acceleration, 9.81	m/s^2
h_1	water depth above the gate top	m
R	resultant force on the gate	N
ρ	water density	kg/m^3

$$
\begin{aligned}
R &= A\rho g \left(\frac{h_1 + h_2}{2} \right) \\
&= (0.20 \text{ m}^2)\left(1000 \ \frac{\text{kg}}{\text{m}^3} \right)\left(9.81 \ \frac{\text{m}}{\text{s}^2} \right) \\
&\quad \times \left(\frac{2.3 \text{ m} + 2.8 \text{ m}}{2} \right) \\
&= 5003 \text{ N}
\end{aligned}
$$

For a circle,

d_R	distance below water surface where the resultant force is applied	m
I_c	centroid area moment of inertia	m^4

$$
\begin{aligned}
I_c &= \frac{\pi D^4}{64} = \frac{\pi (0.5 \text{ m})^4}{64} \\
&= 0.0031 \text{ m}^4
\end{aligned}
$$

$$
\begin{aligned}
d_R &= \frac{h_1 + h_2}{2} + \frac{I_c}{A\left(\dfrac{h_1 + h_2}{2} \right)} \\
&= \frac{2.3 \text{ m} + 2.8 \text{ m}}{2} + \frac{0.0031 \text{ m}^4}{(0.20 \text{ m}^2)\left(\dfrac{2.3 \text{ m} + 2.8 \text{ m}}{2} \right)} \\
&= 2.6 \text{ m}
\end{aligned}
$$

| F_G | force at the non-hinged side of the gate | kN |

$$
\begin{aligned}
F_G &= \frac{R(d_R - h_1)}{D} \\
&= \frac{\left(\dfrac{5003 \text{ N}}{1000 \ \dfrac{\text{N}}{\text{kN}}} \right)(2.6 \text{ m} - 2.3 \text{ m})}{0.5 \text{ m}} \\
&= 3.0 \text{ kN}
\end{aligned}
$$

The answer is (A).

Author Commentary

🕐 Before beginning any calculations, determine what equations are used in the solution, organize them in sequence, and identify values of known terms. This will help proceed to a solution in a logical sequence.

💣 Make sure to calculate the location on the gate where the resultant force is applied rather than using the simple average water depth between the top and bottom of the gate.

42. Find the cross-sectional areas of the pipe and the nozzle.

A_1	pipe cross-sectional area	m^2
A_2	nozzle cross-sectional area	m^2
D_1	pipe diameter	m
D_2	nozzle diameter	m

$$A_1 = \frac{\pi D_1^2}{4} = \frac{\pi \left(\dfrac{2.54 \text{ cm}}{100 \dfrac{\text{cm}}{\text{m}}} \right)^2}{4}$$
$$= 5.1 \times 10^{-4} \text{ m}^2$$

$$A_2 = \frac{\pi D_2^2}{4} = \frac{\pi \left(\dfrac{0.5 \text{ cm}}{100 \dfrac{\text{cm}}{\text{m}}} \right)^2}{4}$$
$$= 2.0 \times 10^{-5} \text{ m}^2$$

Choose a lower water elevation of 0 m to simplify the calculation by omitting the z_1 term. From water properties tables, the density of water at 20°C is 998.3 kg/m^3.

g	gravitational acceleration, 9.81	m/s^2
p_1	water pressure at lower elevation	kPa
p_2	water pressure at upper elevation	kPa
Q_1	flow rate	m^3/s
Q_2	flow rate	m^3/s
z_1	lower water level elevation	m
z_2	upper water level elevation	m
ρ	water density	kg/m^3

$$\frac{p_1}{\rho g} + z_1 + \frac{Q_1^2}{2gA_1^2} = \frac{p_2}{\rho g} + z_2 + \frac{Q_2^2}{2gA_2^2}$$

$$p_1 = p_2 + z_2 \rho g + \frac{Q_2^2 \rho}{2A_2^2} - \frac{Q_1^2 \rho}{2A_1^2}$$

$$= 7000 \text{ kPa} + \frac{(340 \text{ m})\left(998.3 \dfrac{\text{kg}}{\text{m}^3}\right) \times \left(9.81 \dfrac{\text{m}}{\text{s}^2}\right)}{1000 \dfrac{\text{N}}{\text{m}^2 \cdot \text{kPa}}}$$

$$+ \frac{\left(0.002 \dfrac{\text{m}^3}{\text{s}}\right)^2 \left(998.3 \dfrac{\text{kg}}{\text{m}^3}\right)}{(2)(2.0 \times 10^{-5} \text{ m}^2)^2 \left(1000 \dfrac{\text{N}}{\text{m}^2 \cdot \text{kPa}}\right)}$$

$$- \frac{\left(0.002 \dfrac{\text{m}^3}{\text{s}}\right)^2 \left(998.3 \dfrac{\text{kg}}{\text{m}^3}\right)}{(2)(5.1 \times 10^{-4} \text{ m}^2)^2 \left(1000 \dfrac{\text{N}}{\text{m}^2 \cdot \text{kPa}}\right)}$$

$$= 15\,310 \text{ kPa} \quad (15\,000 \text{ kPa})$$

The answer is (D).

Author Commentary

🕐 Because the pump and water surface are at the same elevation, assuming a lower water surface elevation of 0 m will eliminate one term from the equation and simplify the solution.

💣 The final calculation for pressure is where errors are most likely to occur. The calculation includes complicated units and exponents for the flow and area terms. Be careful to balance the units for flow and area properly.

43. The flow velocity in pipe DC is

A	pipe cross-sectional area	ft^2
Q	flow rate	ft^3/sec
v	flow velocity	ft/sec

$$v_{DC} = \frac{Q_{DC}}{A_{DC}} = \frac{4 \dfrac{\text{ft}^3}{\text{sec}}}{0.35 \text{ ft}^2} = 11.4 \text{ ft/sec}$$

The head loss corresponding to a velocity of 11.4 ft/sec in 1600 ft of 8 in diameter pipe is

D	pipe diameter	ft
f	Darcy friction factor	–
g	gravitational acceleration, 32.2	ft/sec^2
h_f	friction head	ft
L	pipe length	ft

$$h_{f,\mathrm{DC}} = \frac{f_{\mathrm{DC}} L_{\mathrm{DC}} v_{\mathrm{DC}}^2}{2 D_{\mathrm{DC}} g} = \frac{(0.016)(1600\ \mathrm{ft})\left(11.4\ \dfrac{\mathrm{ft}}{\mathrm{sec}}\right)^2}{(2)\left(\dfrac{8\ \mathrm{in}}{12\ \dfrac{\mathrm{in}}{\mathrm{ft}}}\right)\left(32.2\ \dfrac{\mathrm{ft}}{\mathrm{sec}^2}\right)}$$

$$= 77.4917\ \mathrm{ft}$$

Determine the direction of flow.

p/γ	pressure head	ft
z	elevation head	ft

$$\frac{p_{\mathrm{D}}}{\gamma} + z_{\mathrm{D}} + \frac{v_{\mathrm{D}}^2}{2g} = \frac{p_{\mathrm{C}}}{\gamma} + z_{\mathrm{C}} + \frac{v_{\mathrm{C}}^2}{2g} + h_{f,\mathrm{DC}}$$

Assume the pressure in tank D, p_{D}, is 0 and the flow velocity is the same in tank D and at node C ($v_{\mathrm{D}} = v_{\mathrm{C}}$). Therefore,

$$\frac{p_{\mathrm{C}}}{\gamma} = z_{\mathrm{D}} - z_{\mathrm{C}} - h_{f,\mathrm{DC}}$$

$$= 676\ \mathrm{ft} - 523\ \mathrm{ft} - 77.4917\ \mathrm{ft}$$
$$= 75.5083\ \mathrm{ft}$$

$$z_{\mathrm{C}} + \frac{p_{\mathrm{C}}}{\gamma} = 523\ \mathrm{ft} + 75.5083\ \mathrm{ft} = 598.5083\ \mathrm{ft}$$

$$z_{\mathrm{A}} = 719\ \mathrm{ft} > 598.5083\ \mathrm{ft}$$
$$z_{\mathrm{D}} = 676\ \mathrm{ft} > 598.5083\ \mathrm{ft}$$

Therefore, flow is from A to C and from D to C, so flow is from C to B.

$$z_{\mathrm{A}} = z_{\mathrm{C}} + \frac{p_{\mathrm{C}}}{\gamma} + h_{f,\mathrm{AC}}$$

$$h_{f,\mathrm{AC}} = z_{\mathrm{A}} - z_{\mathrm{C}} - \frac{p_{\mathrm{C}}}{\gamma}$$

$$= 719\ \mathrm{ft} - 523\ \mathrm{ft} - 75.5083\ \mathrm{ft}$$
$$= 120.4917\ \mathrm{ft}$$

Find the flow velocity from A to C.

$$h_{f,\mathrm{AC}} = \frac{f_{\mathrm{AC}} L_{\mathrm{AC}} v_{\mathrm{AC}}^2}{2 D_{\mathrm{AC}} g}$$

$$v_{\mathrm{AC}} = \sqrt{\frac{h_{f,\mathrm{AC}} 2 D_{\mathrm{AC}} g}{f_{\mathrm{AC}} L_{\mathrm{AC}}}}$$

$$= \sqrt{\frac{(120.4917\ \mathrm{ft})(2)\left(\dfrac{8\ \mathrm{in}}{12\ \dfrac{\mathrm{in}}{\mathrm{ft}}}\right)\left(32.2\ \dfrac{\mathrm{ft}}{\mathrm{sec}^2}\right)}{(0.016)(1200\ \mathrm{ft})}}$$

$$= 16.4144\ \mathrm{ft/sec}$$

Find the flow rates from A to C and from C to B.

$$Q_{\mathrm{AC}} = v_{\mathrm{AC}} A_{\mathrm{AC}} = \left(16.4144\ \frac{\mathrm{ft}}{\mathrm{sec}}\right)(0.35\ \mathrm{ft}^2)$$

$$= 5.74504\ \mathrm{ft}^3/\mathrm{sec}$$

$$Q_{\mathrm{CB}} = Q_{\mathrm{AC}} + Q_{\mathrm{DC}} = 5.75\ \frac{\mathrm{ft}^3}{\mathrm{sec}} + 4\ \frac{\mathrm{ft}^3}{\mathrm{sec}}$$

$$= 9.74504\ \mathrm{ft}^3/\mathrm{sec}$$

The flow velocity from C to B is

$$v_{\mathrm{CB}} = \frac{Q_{\mathrm{CB}}}{A_{\mathrm{CB}}} = \frac{9.74504\ \dfrac{\mathrm{ft}^3}{\mathrm{sec}}}{0.55\ \mathrm{ft}^2} = 17.7182\ \mathrm{ft/sec}$$

The water surface elevation in tank B is

$$h_{f,\mathrm{CB}} = \frac{f_{\mathrm{CB}} L_{\mathrm{CB}} v_{\mathrm{CB}}^2}{2 D_{\mathrm{CB}} g} = \frac{(0.014)(1090\ \mathrm{ft})\left(17.7182\ \dfrac{\mathrm{ft}}{\mathrm{sec}}\right)^2}{(2)\left(\dfrac{10\ \mathrm{in}}{12\ \dfrac{\mathrm{in}}{\mathrm{ft}}}\right)\left(32.2\ \dfrac{\mathrm{ft}}{\mathrm{sec}^2}\right)}$$

$$= 89.2666\ \mathrm{ft}$$

Assume the pressure in tank B, p_{B}, is 0 and the flow velocity is the same in tank B and at node C ($v_{\mathrm{B}} = v_{\mathrm{C}}$). Therefore,

$$\frac{p_{\mathrm{B}}}{\gamma} + z_{\mathrm{B}} + \frac{v_{\mathrm{B}}^2}{2g} = \frac{p_{\mathrm{C}}}{\gamma} + z_{\mathrm{C}} + \frac{v_{\mathrm{C}}^2}{2g} + h_{f,\mathrm{CB}}$$

$$z_{\mathrm{B}} = z_{\mathrm{C}} + \frac{p_{\mathrm{C}}}{\gamma} - h_{f,\mathrm{CB}}$$

$$= 523\ \mathrm{ft} + 75.5083\ \mathrm{ft} - 89.2666\ \mathrm{ft}$$
$$= 509.2417\ \mathrm{ft} \quad (510\ \mathrm{ft})$$

The answer is (B).

44. Steel pipe is typically assumed to be elastic, so correct the bulk modulus of elasticity for use in the speed of sound in the pipe calculation.

From steel pipe dimension tables, 200 mm schedule-40 steel pipe has an inside diameter of 202.7 mm and a wall thickness of 8.18 mm.

D_{pipe}	pipe inside diameter	mm
E	corrected bulk modulus of elasticity	N/m^2
E_{pipe}	steel pipe modulus of elasticity	N/m^2
E_{water}	water bulk modulus of elasticity	N/m^2
t_{pipe}	pipe wall thickness	mm

$$E = \frac{E_{water}t_{pipe}E_{pipe}}{t_{pipe}E_{pipe} + D_{pipe}E_{water}}$$

$$= \frac{\left(2.07 \times 10^9 \ \frac{N}{m^2}\right)(8.18 \text{ mm}) \times \left(200 \times 10^9 \ \frac{N}{m^2}\right)}{(8.18 \text{ mm})\left(200 \times 10^9 \ \frac{N}{m^2}\right) + (202.7 \text{ mm})\left(2.07 \times 10^9 \ \frac{N}{m^2}\right)}$$

$$= 1.6 \times 10^9 \text{ N/m}^2$$

The density of water is 1000 kg/m³. The speed of sound in the pipe is

a	speed of sound in the pipe	m/s
ρ	water density	kg/m³

$$a = \sqrt{\frac{E}{\rho}} = \sqrt{\frac{1.6 \times 10^9 \ \frac{N}{m^2}}{1000 \ \frac{kg}{m^3}}} = 1.265 \times 10^3 \text{ m/s}$$

Assume the pressure instantaneously increases from zero and the velocity instantaneously decreases from 2.4 m/s to zero.

Δp	instantaneous pressure increase in pipe	N/m^2
Δv	instantaneous water velocity decrease	m/s

$$\Delta p = \rho a \Delta v = \left(1000 \ \frac{kg}{m^3}\right)\left(1.265 \times 10^3 \ \frac{m}{s}\right)\left(2.4 \ \frac{m}{s}\right)$$

$$= 3.036 \times 10^6 \text{ N/m}^2$$

g	gravitational acceleration, 9.81	m/s²
$h_{w,max}$	maximum head from water hammer	m

$$h_{w,max} = \frac{\Delta p}{\rho g} = \frac{3.036 \times 10^6 \ \frac{N}{m^2}}{\left(1000 \ \frac{kg}{m^3}\right)\left(9.81 \ \frac{m}{s^2}\right)} = 309 \text{ m} \quad (310 \text{ m})$$

The answer is (B).

Author Commentary

🕐 The values of the modulus of elasticity for water and the modulus of elasticity for pipe material are available from standard tables. Finding these values rather than calculating them saves time.

💣 Using the value of the bulk modulus for water instead of the corrected bulk modulus results in a value much larger than the true value.

45. Find the head loss in the straight pipe due to friction.

h_d	downstream manometer reading	cm
h_f	friction head loss for straight pipe	m/m
h_u	upstream manometer reading	cm
l	pipe length between manometer taps	cm

$$h_f = \frac{h_u - h_d}{l}$$

$$= \frac{14.2 \text{ cm} - 12.9 \text{ cm}}{10 \text{ cm}}$$

$$= 0.13 \text{ cm/cm} \quad (0.13 \text{ m/m})$$

h_m total head loss for straight pipe and   m
    valve

$$h_m = h_u - h_d$$
$$= \frac{18 \text{ cm} - 3.7 \text{ cm}}{100 \dfrac{\text{cm}}{\text{m}}}$$
$$= 0.143 \text{ m}$$

h_v minor loss through valve   m
L pipe length   m

$$h_v = h_m - L h_f$$
$$= 0.143 \text{ m} - \frac{(2)(20 \text{ cm})\left(0.13 \dfrac{\text{m}}{\text{m}}\right)}{100 \dfrac{\text{cm}}{\text{m}}}$$
$$= 0.091 \text{ m}$$

g gravitational acceleration, 9.81   m/s^2
k loss coefficient   –
v flow velocity   m/s

$$h_v = \frac{k \text{v}^2}{2g}$$

$$k = \frac{h_v 2g}{\text{v}^2} = \frac{(0.091 \text{ m})(2)\left(9.81 \dfrac{\text{m}}{\text{s}^2}\right)}{\left(3.2 \dfrac{\text{m}}{\text{s}}\right)^2}$$
$$= 0.17$$

The answer is (A).

Author Commentary

🕐 The difference between the straight pipe head loss and the straight pipe with valve head loss is the head loss through the valve. This simple difference is the independent variable in the loss coefficient equation.

💣 Be sure to include the head loss for the straight pipe as well as for the valve.

46. From water properties tables, the kinematic viscosity of water at 70°F is 1.0×10^{-5} ft^2/sec.

D diameter   in
Q flow rate   ft^3/sec
Re Reynolds number   –
ν kinematic viscosity   ft^2/sec

$$\text{Re} = \frac{4Q}{\pi D \nu} = \frac{(4)\left(950 \dfrac{\text{gal}}{\text{min}}\right)\left(12 \dfrac{\text{in}}{\text{ft}}\right)}{\pi(10 \text{ in})\left(1.0 \times 10^{-5} \dfrac{\text{ft}^2}{\text{sec}}\right)\left(7.48 \dfrac{\text{gal}}{\text{ft}^3}\right)} \times \left(60 \dfrac{\text{sec}}{\text{min}}\right)$$
$$= 3.242 \times 10^5$$

The relative roughness is

ϵ specific roughness   in
ϵ/D relative roughness   –

$$\frac{\epsilon}{D} = \frac{(0.000008 \text{ ft})\left(12 \dfrac{\text{in}}{\text{ft}}\right)}{10 \text{ in}} = 0.0000096$$

From the Moody friction factor chart, for a Reynolds number of 3.242×10^5 and a relative roughness of 0.0000096, the Darcy friction factor is 0.014.

f Darcy friction factor   –
g gravitational acceleration, 32.2   ft/sec^2
h_f head loss due to friction   ft
L length   ft

$$h_f = \frac{8 f L Q^2}{\pi^2 g D^5}$$

$$= \frac{(8)(0.014)(1000 \text{ ft})\left(\dfrac{950 \dfrac{\text{gal}}{\text{min}}}{\left(7.48 \dfrac{\text{gal}}{\text{ft}^3}\right)\left(60 \dfrac{\text{sec}}{\text{min}}\right)}\right)^2}{\pi^2\left(32.2 \dfrac{\text{ft}}{\text{sec}^2}\right)\left(\dfrac{10 \text{ in}}{12 \dfrac{\text{in}}{\text{ft}}}\right)^5}$$

$$= 3.95 \text{ ft} \quad (4.0 \text{ ft})$$

The answer is (C).

Author Commentary

🕐 This is a Darcy equation problem for which there is a direct solution. Proceed sequentially from the Reynolds number calculation, to the relative roughness calculation, to the friction factor calculation. Use tables or the Moody friction factor chart to determine the friction factor.

💣 All units of length and time should be in feet and seconds, respectively, throughout the calculations.

47. The customary U.S. discharge equation for a Cipoletti weir is

b	weir width	ft
H	head above notch	ft
Q	flow rate	ft³/sec

$$Q = 3.367bH^{3/2}$$

$$= (3.367)\left(\frac{24 \text{ in}}{12 \frac{\text{in}}{\text{ft}}}\right)\left(\frac{8.6 \text{ in}}{12 \frac{\text{in}}{\text{ft}}}\right)^{3/2}$$

$$= 4.1 \text{ ft}^3/\text{sec}$$

The answer is (B).

Author Commentary

🕐 This is a simple one-step calculation, but it requires selecting the proper equation. Use the customary U.S. discharge equation for a Cipoletti weir.

48. For a 1:1 side-wall slope, the side-wall angle is 45°.

b	channel base width	m
d	water depth	m
R	hydraulic radius	m
θ	side-wall angle measured from the horizontal	deg

$$R = \frac{bd \sin \theta + d^2 \cos \theta}{b \sin \theta + 2d}$$

$$= \frac{b(1 \text{ m})(\sin 45°) + (1 \text{ m})^2(\cos 45°)}{b(\sin 45°) + (2)(1 \text{ m})}$$

$$= \frac{0.707b + 0.707}{0.707b + 2} \text{ m}$$

Because the side walls and base are constructed of different materials, use an average Manning roughness coefficient. The roughness coefficient for concrete is 0.013. The roughness coefficient for firm gravel is 0.023. The average Manning roughness coefficient is

| n | Manning roughness coefficient | — |

$$n_{\text{ave}} = \frac{0.013 + 0.023}{2}$$

$$= 0.018$$

| S | channel slope | m/m |
| v | flow velocity | m/s |

$$\text{v} = \left(\frac{1}{n}\right)R^{2/3}\sqrt{S}$$

$$0.75 \frac{\text{m}}{\text{s}} = \left(\frac{1}{0.018}\right)\left(\frac{0.707b + 0.707}{0.707b + 2} \text{ m}\right)^{2/3}$$
$$\times \sqrt{0.0005 \frac{\text{m}}{\text{m}}}$$

$$\left(0.604 \frac{\text{m}}{\text{s}}\right)^{3/2} = \frac{0.707b + 0.707}{0.707b + 2} \text{ m}$$

$$\left(0.469 \frac{\text{m}}{\text{s}}\right) = 0.707b + 0.707 \text{ m}$$
$$\times (0.707b + 2 \text{ m})$$

$$b = 0.62 \text{ m}$$

The answer is (A).

Author Commentary

🕐 Define the hydraulic radius in terms of the channel base. With the Manning equation, this produces a function where the unknown cannot be isolated and requires a solution by trial and error, by numerical methods, or by using a solve function on a calculator. Using a solve function is the quickest method for finding an accurate solution to the problem.

💣 The value of n is very important in this calculation, so make sure to choose n carefully for each channel material and to calculate an average value for n.

49. For 3:1 horizontal-to-vertical slope,

θ — side slope angle measured from the horizontal — deg

$$\theta = \tan^{-1}\left(\frac{1}{3}\right) = 18.4°$$

A — cross-sectional area of channel — m^2
b — base width — m
d — water depth — m

$$A = \left(b + \frac{d}{\tan\theta}\right)d = \left(2 \text{ m} + \frac{1 \text{ m}}{\tan 18.4°}\right)(1 \text{ m})$$
$$= 5.0 \text{ m}^2$$

P — wetted perimeter — m

$$P = b + 2\left(\frac{d}{\sin\theta}\right) = 2 \text{ m} + (2)\left(\frac{1 \text{ m}}{\sin 18.4°}\right)$$
$$= 8.3 \text{ m}$$

S — channel slope — m/m
ΔL — change in distance — m
Δz — change in elevation — m

$$S = \frac{\Delta z}{\Delta L} = \frac{20 \text{ m}}{(9 \text{ km})\left(1000 \dfrac{\text{m}}{\text{km}}\right)} = 0.0022 \text{ m/m}$$

The Manning roughness coefficient for earth is 0.018.

n — Manning roughness coefficient — –
Q — flow rate — m^3/s

$$Q = \frac{\left(\dfrac{1}{n}\right)A^{5/3}\sqrt{S}}{P^{2/3}}$$
$$= \frac{\left(\dfrac{1}{0.018}\right)(5.0 \text{ m}^2)^{5/3}\sqrt{0.0022 \dfrac{\text{m}}{\text{m}}}}{(8.3 \text{ m})^{2/3}}$$
$$= 9.3 \text{ m}^3/\text{s}$$

The answer is (B).

Author Commentary

💣* Be careful with the side slope angle calculation. A 3:1 side slope does not have the same slope angle as a 1:3 side slope.

50. Because the length of a low-lead siphon is relatively short compared to the throat diameter, any head loss through the siphon can be assumed to be negligible.

A_o — throat cross-sectional area — ft^2
C_d — siphon entrance coefficient — –
D — siphon throat diameter — ft
g — gravitational acceleration, 32.2 — ft/sec^2
h — operating head — ft
Q — discharge flow — ft^3/sec

$$A_o = \frac{\pi D^2}{4}$$
$$Q = C_d A_o \sqrt{2gh} = \frac{C_d \pi D^2 \sqrt{2gh}}{4}$$
$$D = \sqrt{\frac{4Q}{C_d \pi \sqrt{2gh}}}$$
$$= \sqrt{\frac{(4)\left(90 \dfrac{\text{ft}^3}{\text{sec}}\right)}{0.6\pi\sqrt{(2)\left(32.2 \dfrac{\text{ft}}{\text{sec}^2}\right)(4 \text{ ft})}}}$$
$$= 3.45 \text{ ft} \quad (3.5 \text{ ft})$$

Because a low-head siphon operates at negative pressure, the limitation of subatmospheric pressure should be checked. The vortex flow equation can be used for this purpose.

R_c — radius to throat centerline — ft
R_s — radius to throat diameter — ft

$$R_s = R_c + \frac{D}{2}$$
$$= 2 \text{ ft} + \frac{3.45 \text{ ft}}{2}$$
$$= 3.725 \text{ ft}$$

p_a	atmospheric pressure head	ft

$$Q \le R_c\sqrt{(0.7)(2gp_a)} \ln\frac{R_s}{R_c}$$
$$= 2 \text{ ft}\sqrt{(0.7)(2)\left(32.2\ \frac{\text{ft}}{\text{sec}^2}\right)(34 \text{ ft})}\ \ln\left(\frac{3.725 \text{ ft}}{2 \text{ ft}}\right)$$
$$= 48.7 \text{ ft}^3/\text{sec} < 90 \text{ ft}^3/\text{sec} \quad [\text{OK}]$$

The answer is (D).

Author Commentary

🕐 For a low-head siphon spillway, the orifice equation arranged with throat diameter as the dependent variable is the quickest solution, as it requires only a single calculation.

💣 Because a low-head siphon operates at negative pressure, the limitation of subatmospheric pressure is usually checked in common practice. However, because this requires additional calculations, skipping this check may save time, although doing so risks an incorrect answer.

51. Find the wetted perimeter of the culvert barrel.

D	culvert barrel diameter	ft
P	culvert barrel wetted perimeter	ft

$$P = \pi D = \pi(6.0 \text{ ft}) = 18.85 \text{ ft}$$

A	culvert barrel cross-sectional area	ft^2

$$A = \frac{\pi D^2}{4} = \frac{\pi(6.0 \text{ ft})^2}{4} = 28.3 \text{ ft}^2$$

R	culvert barrel hydraulic radius	ft

$$R = \frac{A}{P} = \frac{28.3 \text{ ft}^2}{18.85 \text{ ft}} = 1.5 \text{ ft}$$

The Manning roughness coefficient for concrete is 0.013. For square flush openings at the culvert inlet and outlet, the discharge coefficient is assumed to be equal to the orifice coefficient for a sharp-edged opening, 0.62.

C_d	discharge coefficient	—
f_c	friction loss correction	—
L	culvert length	ft
n	Manning roughness coefficient	—

$$f_c = 1 + \frac{29C_d^2 n^2 L}{R^{4/3}} = 1 + \frac{(29)(0.62)^2(0.013)^2(276 \text{ ft})}{(1.5 \text{ ft})^{4/3}}$$
$$= 1.3$$

$d_{\text{CL},i}$	water depth above the culvert centerline at the inlet	ft
d_i	inlet elevation head relative to the outlet invert	ft
S	culvert slope	ft/ft

$$d_i = 0.5D + d_{\text{CL},i} + LS$$
$$= (0.5)(6.0 \text{ ft}) + 12.5 \text{ ft} + (276 \text{ ft})\left(\frac{1 \text{ ft}}{83 \text{ ft}}\right)$$
$$= 18.8 \text{ ft}$$

$d_{\text{CL},o}$	water depth above the culvert centerline at the outlet	ft
d_o	outlet elevation head relative to the outlet invert	ft

$$d_o = 0.5D + d_{\text{CL},o}$$
$$= (0.5)(6.0 \text{ ft}) + 4.6 \text{ ft}$$
$$= 7.6 \text{ ft}$$

For the culvert depicted in the illustration, the flow is submerged outlet flow. The maximum discharge for submerged outlet flow can be approximated as

g	gravitational acceleration, 32.2	ft/sec^2
Q	discharge	ft^3/sec

$$Q = C_d A\sqrt{\frac{2g(d_i - d_o)}{f_c}}$$
$$= (0.62)(28.3 \text{ ft}^2)$$
$$\times \sqrt{\frac{(2)\left(32.2\ \frac{\text{ft}}{\text{sec}^2}\right)(18.8 \text{ ft} - 7.6 \text{ ft})}{1.3}}$$
$$= 413 \text{ ft}^3/\text{sec} \quad (410 \text{ ft}^3/\text{sec})$$

The answer is (B).

Author Commentary

🕐 Don't waste time looking for the discharge coefficient. For square flush openings at the culvert inlet and outlet, the discharge coefficient is equal to the orifice coefficient of a sharp-edged opening.

💣 Be careful when determining the culvert flow classification. Observe the position of the culvert outlet relative to the water surface. This is the purpose of the problem illustration and indicates which culvert equation to use.

52. Rearrange the Manning equation in terms of slope.

n	Manning roughness coefficient	–
R	hydraulic radius	m
S	slope	m/m
v	velocity	m/s

$$\text{v} = \left(\frac{1}{n}\right)R^{2/3}\sqrt{S}$$

$$S = \frac{n^2\text{v}^2}{R^{4/3}} \quad [\text{Eq. 1}]$$

Combine the Darcy equation, the equation for slope as a function of head loss, and the equation relating diameter and hydraulic radius for a circular channel flowing half full.

D	diameter	m
f	Darcy friction factor	–
g	gravitational acceleration, 9.81	m/s²
h_f	head loss due to friction	m
L	length	m

$$h_f = \frac{fL\text{v}^2}{2gD}$$

$$S = \frac{h_f}{L}$$

$$D = 4R$$

$$S = \frac{f\text{v}^2}{2g4R} \quad [\text{Eq. 2}]$$

Set Eq. 1 equal to Eq. 2. Solve for the Darcy friction factor. Units will not cancel.

$$\frac{f\text{v}^2}{2g4R} = \frac{n^2\text{v}^2}{R^{4/3}}$$

$$f = \frac{n^2 8g}{R^{1/3}} = \frac{(0.013)^2(8)\left(9.81\ \dfrac{\text{m}}{\text{s}^2}\right)}{(1\ \text{m})^{1/3}}$$

$$= 0.013$$

The answer is (B).

Author Commentary

🕐 Find a common variable between the Darcy and Manning equations, set the equations equal to each other, and rearrange terms to isolate the friction factor. Once the friction factor equation is developed, there is only a single, simple calculation required to find the answer.

💣 The Manning coefficient has units, but is typically assumed to be unitless. Because of this, the units in the calculation will not cancel properly. Use consistent base units (meters and seconds) to arrive at the correct answer.

53. From the problem illustration, at 12 h the discharge is 34 m³/s, and the gaging station measures 150 m³/s. The unit hydrograph shows the peak occurring at 6 h with a discharge of 59 m³/s.

Q	peak discharge	m³/s
Q_{12}	hydrograph discharge at 12 h	m³/s
Q_h	hydrograph peak discharge	m³/s
Q_o	measured peak discharge	m³/s

$$\frac{Q}{Q_h} = \frac{Q_o}{Q_{12}}$$

$$Q = \frac{Q_h Q_o}{Q_{12}}$$

$$= \frac{\left(59\ \dfrac{\text{m}^3}{\text{s}}\right)\left(150\ \dfrac{\text{m}^3}{\text{s}}\right)}{34\ \dfrac{\text{m}^3}{\text{s}}}$$

$$= 260\ \text{m}^3/\text{s}$$

The answer is (D).

Author Commentary

🕐 The quickest solution to this problem is to use a simple ratio of flows. One flow is given, and the other two are taken from the unit hydrograph. This leaves the peak discharge unknown.

💣※ Although the solution involves using simple ratios, be careful selecting the proper ratios. Use common time for one ratio and peak discharges for the other.

54. Calculate the net area. For example, for area II, $252 \text{ km}^2 - 84 \text{ km}^2 = 168 \text{ km}^2$.

Next, determine the average precipitation, which is taken between two adjacent isohyets. For example, for area II,

$$\frac{20 \text{ cm} + 22 \text{ cm}}{2} = 21 \text{ cm}$$

For areas I and VI, which are not adjacent, use the pattern of increasing or decreasing precipitation values to estimate the precipitation. Area I must be greater than 22 cm, so use 23 cm. Likewise, area VI must be less than 14 cm, so use 13 cm.

The precipitation volume for each area is the product of the net area and the average precipitation. The precipitation volumes for all areas are summed to determine the cumulative precipitation volume.

Tabulate the values for each area as shown.

area	isohyet (cm)	enclosed area (km^2)	net area (km^2)	average precipitation (cm)	precipitation volume $(\text{cm}\cdot\text{km}^2)$
I	>22	84	84	23	1932
II	20	252	168	21	3528
III	18	578	326	19	6194
IV	16	892	314	17	5338
V	14	1136	244	15	3660
VI	<14	1294	158	13	2054
					22 706

Calculate the average precipitation of the region.

A	total area	km^2
P_m	average precipitation	cm
P	cumulative precipitation volume	$\text{cm}\cdot\text{km}^2$

$$P_m = \frac{\sum P}{A} = \frac{22\,706 \text{ cm}\cdot\text{km}^2}{1294 \text{ km}^2}$$
$$= 17.5 \text{ cm} \quad (18 \text{ cm})$$

The answer is (C).

Author Commentary

🕐 The cumulative annual precipitation volume must be calculated for each net area, then the values for each net area must be summed. To save time, use the three columns in the problem statement as the first three columns of a solution table.

💣※ The precipitation shown in the illustration is at the boundary. To find the cumulative annual precipitation volume for each area, it is necessary to use the average precipitation between the two boundaries. If the average isn't used, there won't be a discrete volume for each net area, and a solution will not be possible.

55. From the problem illustration, 40 recorded events experienced a flow exceeding 40,000 ft^3/sec.

m	number of events of interest that exceed peak flow	—
N	period of record	yr
t_p	recurrence interval	yr

$$t_p = \frac{N+1}{m} = \frac{112 \text{ yr} + 1}{40}$$
$$= 2.8 \text{ yr}$$

The answer is (C).

56. Because the watershed is small and rainfall intensity, ground surface slope, and flow distance are known, the appropriate equation to calculate time of concentration would be either the Izzard equation or the kinematic wave formula. Use the kinematic wave equation.

For manicured sod, the Manning roughness coefficient for overland flow is approximately 0.30.

n Manning roughness coefficient for   –
     overland flow

S ground surface slope   ft/ft

t_c time of concentration   min

$$t_c = \frac{0.94 L^{0.6} n^{0.6}}{i^{0.4} S^{0.3}} = \frac{(0.94)(270 \text{ ft})^{0.6}(0.30)^{0.6}}{\left(2.1 \, \frac{\text{in}}{\text{hr}}\right)^{0.4}\left(0.011 \, \frac{\text{ft}}{\text{ft}}\right)^{0.3}}$$

$$= 37.8 \text{ min} \quad (38 \text{ min})$$

The answer is (D).

Author Commentary

🕐   Several equations are available for calculating time of concentration. Use the given information to select the most appropriate equation. Use the simplest equation if more than one is suitable.

💣   Some equations are specific to SI or customary U.S. units. Confirm which kind of units are required for the equation and do all necessary conversions.

57. The elevation head is

h elevation head above the orifice   ft

$$h = 100 \text{ ft} - 92.6 \text{ ft}$$
$$= 7.4 \text{ ft}$$

For a sharp-edged orifice, the orifice coefficient is 0.62.

A orifice area ft^2
C_d orifice coefficient   –
g gravitational acceleration, 32.2 ft/sec^2
Q discharge rate through the orifice ft^3/sec

$$A = \frac{Q}{C_d \sqrt{2gh}} = \frac{20 \, \frac{\text{ft}^3}{\text{sec}}}{0.62\sqrt{(2)\left(32.2 \, \frac{\text{ft}}{\text{sec}^2}\right)(7.4 \text{ ft})}}$$

$$= 1.48 \text{ ft}^2$$

D orifice diameter   ft

$$D = \sqrt{\frac{4A}{\pi}} = \sqrt{\frac{(4)(1.48 \text{ ft}^2)}{\pi}}$$
$$= 1.4 \text{ ft}$$

The answer is (D).

58. Use a log Pearson type III distribution to find the flood magnitude.

N number of years of record   –
$\log X$ log of the flood magnitude   –
$\log \overline{X}$ mean of log value of all floods   –
σ standard deviation   –

$$\sigma = \sqrt{\frac{\sum(\log X - \log \overline{X})^2}{N-1}} = \sqrt{\frac{3.894}{97-1}} = 0.20$$

g skew coefficient   –

$$g = \frac{N\sum(\log X - \log \overline{X})^3}{(N-1)(N-2)\sigma^3}$$
$$= \frac{(97)(0.181)}{(97-1)(97-2)(0.20)^3}$$
$$= 0.24$$

From a standard reference table for the log Pearson type III distribution with a recurrence interval of 25 yr and a skew coefficient of 0.24, the log Pearson type III distribution coefficient is

K log Pearson type III distribution   –
     coefficient
X flood magnitude ft^3/sec

$$K = 1.833$$
$$\log X = \log \overline{X} + K\sigma = 3.571 + (1.833)(0.20)$$
$$= 3.94$$
$$X = 8662 \text{ ft}^3/\text{sec} \quad (8700 \text{ ft}^3/\text{sec})$$

The answer is (C).

Author Commentary

🕓 Because the problem statement describes the flood using specific statistical functions, it is necessary to find a statistical distribution that uses these functions. The log Pearson type III distribution uses these functions, and recognizing this will save time and lead to a straightforward solution. For problems like this one, use the given information as a hint to which equations apply.

59. Find the effective diffusion.

D_d	diffusion coefficient	m^2/s
D^*	effective diffusion	m^2/s
ω	tortuosity	–

$$D^* = \omega D_d = (0.6)\left(8.7 \times 10^{-9} \ \frac{m^2}{s}\right)$$
$$= 5.2 \times 10^{-9} \ m^2/s$$

C	concentration of the solute at time t	mg/L
C_0	concentration of the solute at time zero	mg/L
erfc	complementary error function	–
t	travel time of interest	d
x	liner thickness	m

$$\frac{C}{C_0} = \text{erfc}\frac{x}{2\sqrt{D^*t}} = \frac{100 \ \frac{mg}{L}}{12\,000 \ \frac{mg}{L}} = 0.00833$$

Using complementary error function tables,

$$\frac{x}{2\sqrt{D^*t}} = 1.87$$
$$x = (1.87)(2)\sqrt{D^*t}$$
$$= (1.87)(2)\sqrt{\left(5.2 \times 10^{-9} \ \frac{m^2}{s}\right)(100 \ yr) \times \left(365 \ \frac{d}{yr}\right)\left(86\,400 \ \frac{s}{d}\right)}$$
$$= 15 \ m$$

The answer is (B).

Author Commentary

🕓 The transport of the tracer is diffusion-dominated, so the dispersion terms in the calculation can be ignored.

💣✳ The complementary error function, erfc, can be confusing to those unfamiliar with it. Be careful when using the erfc tables.

60. In the problem illustration, the flow lines show the path along which an ideal water particle travels on its way to the extraction wells pumping at steady state. The tick marks show the distance traveled by an ideal water particle in uniform time increments. The time increment is identified as 10 days. Where the tick marks are spaced very closely, the water travels a very short distance during the 10 day time increment. In the area between two extraction wells, the water particle is being acted upon by each well and effectively becomes suspended between the two. This represents stagnation and is shown in the illustration at circled letter A. At B and C, the tick mark spacing is great enough to show that the water is flowing toward the wells—not indicative of stagnation. Circled letter D indicates an extraction well, the point of highest groundwater velocity.

The answer is (A).

Author Commentary

🕓 Examine the illustration and read the boxed caption. Stagnation occurs where tick marks are very closely spaced. This only occurs at A. There is no need to examine the other points.

61. The hydraulic conductivity of the soil is

a	auger hole diameter	ft
b	distance between auger holes	ft
K	hydraulic conductivity	ft/day
L	aquifer thickness	ft
Q	pumping rate	ft^3/day
Δh	head difference between auger holes	ft

$$K = \frac{Q}{\pi L \Delta h} \cosh^{-1} \frac{b}{2a}$$

$$= \frac{\left(26 \frac{\text{gal}}{\text{min}}\right)\left(1440 \frac{\text{min}}{\text{day}}\right)}{\pi(15 \text{ ft})\left(\dfrac{9.7 \text{ in}}{12 \dfrac{\text{in}}{\text{ft}}}\right)\left(7.48 \dfrac{\text{gal}}{\text{ft}^3}\right)}$$

$$\times \cosh^{-1} \frac{10 \text{ ft}}{(2)\left(\dfrac{8 \text{ in}}{12 \dfrac{\text{in}}{\text{ft}}}\right)}$$

$$= 356 \text{ ft/day} \quad (360 \text{ ft/day})$$

The answer is (C).

Author Commentary

🕐 Look for an equation describing the two-auger-hole method for determining hydraulic conductivity. This equation requires a single calculation, and all required information is given. Finding the equation is the time-consuming step.

💣 This is a straightforward calculation. It requires care because it involves conversion factors and an uncommon trigonometric function.

62. Find the horizontal velocity.

d_p	mean particle diameter	m
f	Darcy friction factor	–
g	gravitational acceleration, 9.81	m/s^2
k	Camp formula constant	–
SG_p	grit specific gravity	–
v	horizontal velocity	ft/sec or m/s

$$\text{v} = \sqrt{\frac{8kgd_p(SG_p - 1)}{f}}$$

$$= \sqrt{\frac{(8)(0.05)\left(9.81 \dfrac{\text{m}}{\text{s}^2}\right)(0.22 \text{ mm})(2.65 - 1)}{(0.03)\left(1000 \dfrac{\text{mm}}{\text{m}}\right)}}$$

$$= 0.22 \text{ m/s}$$

A	channel cross-sectional area	ft^2
d	depth	ft
Q	flow rate	ft^3/sec
w	width	ft

$$Q = A\,\text{v} = wd\text{v}$$

$$w = \frac{Q}{d\text{v}}$$

$$= \frac{3.5 \times 10^6 \dfrac{\text{gal}}{\text{day}}}{(4 \text{ ft})\left(0.22 \dfrac{\text{m}}{\text{s}}\right)\left(3.28 \dfrac{\text{ft}}{\text{m}}\right)\left(86{,}400 \dfrac{\text{sec}}{\text{day}}\right)\left(7.48 \dfrac{\text{gal}}{\text{ft}^3}\right)}$$

$$= 1.9 \text{ ft}$$

The answer is (B).

Author Commentary

🕐 The Camp formula constant and Darcy friction factor are given in the problem statement, while velocity, which is required to calculate the basin width, is not. Save time by finding an equation that uses these parameters to find velocity.

💣 Once the appropriate equation is selected, the calculations are simple. Errors will most likely occur in unit conversions.

63. The sludge volume index is

MLSS	mixed liquor suspended solids concentration	mg/L
SV	sludge volume	mL/L
SVI	sludge volume index	mL/g

$$SVI = \frac{SV}{MLSS}$$

$$= \frac{\left(356 \dfrac{\text{mL}}{\text{L}}\right)\left(1000 \dfrac{\text{mg}}{\text{g}}\right)}{2400 \dfrac{\text{mg}}{\text{L}}}$$

$$= 148 \text{ mL/g} \quad (150 \text{ mL/g})$$

The answer is (A).

Author Commentary

🕐 If the definition of a sludge volume index is known, this is an easy problem involving a simple ratio. If sludge volume indexes are unfamiliar, examine the units given in the problem statement and those used in the answer options. Organizing the given information to produce the required units will lead to the correct answer.

64. Find the maximum growth rate for denitrification.

k	growth rate	d^{-1}
Y	yield coefficient	g/g
μ_{mD}	maximum growth rate fordenitrification	d^{-1}

$$\mu_{mD} = kY$$
$$= \left(\frac{0.38}{d}\right)\left(0.81 \ \frac{g}{g}\right)$$
$$= 0.31 \ d^{-1}$$

C_N	nitrate and/or nitrite concentration	mg/L

$$C_N = 29 \ \frac{mg}{L} + 8 \ \frac{mg}{L} = 37 \ mg/L$$

C_M	methanol concentration	mg/L
K_M	half-velocity constant for methanol	mg/L
K_N	half-velocity constant for nitrogen	mg/L
T	temperature	°C
μ'_{mD}	corrected maximum growth rate for denitrification	d^{-1}

$$\mu'_{mD} = \mu_{mD}0.0025 \, T^2\left(\frac{C_M}{K_M + C_M}\right)\left(\frac{C_N}{K_N + C_N}\right)$$
$$= \left(\frac{0.31}{d}\right)(0.0025)(16°C)^2\left(\frac{72 \ \frac{mg}{L}}{12 \ \frac{mg}{L} + 72 \ \frac{mg}{L}}\right)$$
$$\times \left(\frac{37 \ \frac{mg}{L}}{0.31 \ \frac{mg}{L} + 37 \ \frac{mg}{L}}\right)$$
$$= 0.17 \ d^{-1}$$

The answer is (C).

Author Commentary

🕐 Find the equation for the corrected maximum growth rate for nitrification from the equations for growth rate, maximum growth rate, and corrected maximum growth rate. This equation is sometimes given in parts. Instead of addressing corrections for temperature, methanol, and nitrate separately, save time by using an equation that combines all three.

💣 Several terms used in the solution are represented by similar symbols, such as the growth rate and the half-velocity constants. Concentrations and half-velocity constants are also represented by identical symbols, differentiated by their subscripts. Don't confuse these terms.

65. The design criteria are based on population equivalents (PE). A PE represents the equivalent organic loading from one person. Each community member constitutes a complete population equivalent.

A	surface area	m^2

$$A = (800 \ people)\left(1 \ \frac{PE}{person}\right)\left(6.0 \ \frac{m^2}{PE}\right)$$
$$= 4800 \ m^2$$

d	bed depth	m
V	empty-bed volume	m^3

$$V = Ad = (4800 \ m^2)(0.75 \ m)$$
$$= 3600 \ m^3$$

Q	flow rate	m^3/d

$$Q = (800 \ people)\left(1 \ \frac{PE}{person}\right)\left(0.2 \ \frac{m^3}{PE \cdot d}\right)$$
$$= 160 \ m^3/d$$

t empty-bed hydraulic residence time d

$$t = \frac{V}{Q} = \frac{3600 \text{ m}^3}{160 \ \dfrac{\text{m}^3}{\text{d}}}$$

$$= 22.5 \text{ d} \quad (23 \text{ d})$$

The answer is (C).

Author Commentary

🕐 Both volume and flow rate are calculated from the population equivalent (PE). Once the definition of PE is known, the solution will follow quickly, so checking the definition will speed up the solution.

💣 Standard equations do not exist for all of the calculations in this solution. Keep track of units and cancel those that are common to find the correct solution.

---

66. Assume first-order kinetics describe the bacteria population die-off.

C concentration at time of interest %
C_o initial concentration %
k first-order rate coefficient d^{-1}
t time d

$$t = \frac{-\ln\dfrac{C}{C_o}}{k} = \frac{-\ln\left(\dfrac{100\% - 90\%}{100\%}\right)}{\dfrac{1.0}{\text{d}}}$$

$$= 2.3 \text{ d}$$

The answer is (C).

Author Commentary

🕐 Bacterial die-off follows first-order kinetics. Knowing this leads directly to the equation for first-order kinetics.

💣 Do not calculate die-off for 10% instead of 90% or calculate the half-life.

---

67. Solve for the loading rate.

k loading rate $\text{L/m}^2\cdot\text{d}$
n number of stages –
S_i initial (influent) 5-day BOD mg/L
S_o final (effluent) 5-day BOD mg/L

$$\frac{S_o}{S_i} = \left(\frac{1}{1 + \dfrac{k_{\text{design}}}{k}}\right)^n$$

$$\frac{40 \ \dfrac{\text{mg}}{\text{L}}}{176 \ \dfrac{\text{mg}}{\text{L}}} = \left(\frac{1}{1 + \dfrac{40.6 \ \dfrac{\text{L}}{\text{m}^2\cdot\text{d}}}{k}}\right)^3$$

$$k = 63.6 \text{ L/m}^2\cdot\text{d}$$

The loading rate is equal to Q/A, so rearrange to solve for the area of each stage, then determine the total area.

A area m^2
Q flow rate gal/d

$$k = \frac{Q}{A_{\text{stage}}}$$

$$A_{\text{stage}} = \frac{Q}{k} = \frac{\left(0.5 \times 10^6 \ \dfrac{\text{gal}}{\text{d}}\right)\left(3.785 \ \dfrac{\text{L}}{\text{gal}}\right)}{63.6 \ \dfrac{\text{L}}{\text{m}^2\cdot\text{d}}} = 29\,756 \text{ m}^2$$

$$A_{\text{total}} = (3)(29\,756 \text{ m}^2) = 89\,269 \text{ m}^2 \quad (89\,000 \text{ m}^2)$$

The answer is (C).

Author Commentary

🕐 Rotating biological contactor (RBC) design uses empirical equations with specific units. Find these equations in relevant references, and convert all parameters to the units defined for the equations. This will save time and reduce the chance of mistakes.

💣 The solution asks for the total disc surface area. The empirical equations may give only area per stage, so be sure that the final step includes a calculation for the total area.

68. The volatile suspended solids concentration is

MSI	mass of ignited crucible, filter paper, and solids	g
MSS	mass of dried crucible, filter paper, and solids	g
VF	sample volume filtered	mL
VSS	volatile suspended solids	mg/L

$$
\begin{aligned}
\text{VSS} &= \frac{\text{MSS} - \text{MSI}}{\text{VF}} \\
&= \left(\frac{25.645 \text{ g} - 25.501 \text{ g}}{200 \text{ mL}} \right) \left(10^6 \ \frac{\text{mL·mg}}{\text{L·g}} \right) \\
&= 720 \text{ mg/L}
\end{aligned}
$$

The answer is (C).

Author Commentary

🕐 Save time by applying the conversion factor once for the whole calculation instead of converting each term individually.

💣 The most common mistake in determining the VSS concentration is using the mass of the ignited solids instead of the ash that remains.

69. Because they are routinely cultured in the laboratory, it is desirable that indicator organisms not be pathogenic. Although some coliform organisms are infectious, they are generally regarded as nonpathogenic.

The answer is (D).

Author Commentary

🕐 Consult a reference to solve this problem, but save time by first considering which characteristic would be undesirable for an indicator organism.

70. The EPA-recommended exposure factors are

A	absorbed	–
BW	body weight	kg
C	concentration	mg/L
DI	daily intake	L/d
ED	exposed duration	yr
LT	lifetime	yr
T	time exposed	–

A	1
BW	70 kg
C	100 μg/L
DI	2 L/d
ED	30 yr
LT	70 yr
T	1

Use a 30-year duration for exposure and a 70 kg adult body mass. For trichloroethylene (TCE),

CDI	chronic daily intake	mg/kg·d

$$
\begin{aligned}
C &= \left(100 \ \frac{\mu\text{g}}{\text{L}} \right) \left(0.001 \ \frac{\text{mg}}{\mu\text{g}} \right) \\
&= 0.1 \text{ mg/L}
\end{aligned}
$$

$$
\begin{aligned}
\text{CDI} &= \frac{C(\text{DI})\,A(\text{ED})\,T}{(\text{BW})(\text{LT})} \\
&= \frac{\left(0.1 \ \frac{\text{mg}}{\text{L}} \right) \left(2 \ \frac{\text{L}}{\text{d}} \right)(1)(30 \text{ yr})(1)}{(70 \text{ kg})(70 \text{ yr})} \\
&= 0.00122 \text{ mg/kg·d}
\end{aligned}
$$

The lifetime risk is

PF	potency factor	$(\text{mg/kg·d})^{-1}$

$$
\begin{aligned}
(\text{CDI})(\text{PF}_{\text{TCE}}) &= \left(0.00122 \ \frac{\text{mg}}{\text{kg·d}} \right) \left(\frac{0.011}{\frac{\text{mg}}{\text{kg·d}}} \right) \\
&= 1.3 \times 10^{-5}
\end{aligned}
$$

For 1,1-dichloroethylene (1,1-DCE),

$$
\begin{aligned}
C &= \left(7.2 \ \frac{\mu\text{g}}{\text{L}} \right) \left(0.001 \ \frac{\text{mg}}{\mu\text{g}} \right) \\
&= 0.0072 \text{ mg/L}
\end{aligned}
$$

$$
\begin{aligned}
\text{CDI} &= \frac{C(\text{DI})\,A(\text{ED})\,T}{(\text{BW})(\text{LT})} \\
&= \frac{\left(0.0072 \ \frac{\text{mg}}{\text{L}} \right) \left(2 \ \frac{\text{L}}{\text{d}} \right)(1)(30 \text{ yr})(1)}{(70 \text{ kg})(70 \text{ yr})} \\
&= 0.000088 \text{ mg/kg·d}
\end{aligned}
$$

The lifetime risk is

$$(CDI)(PF_{1,1\text{-DCE}}) = \left(0.000088 \; \frac{mg}{kg \cdot d}\right)\left(\frac{0.58}{\frac{mg}{kg \cdot d}}\right)$$
$$= 5.1 \times 10^{-5}$$

The total lifetime risk is

$$(1.3 \times 10^{-5}) + (5.1 \times 10^{-5}) = 6.4 \times 10^{-5}$$
$$\text{(64 in one million)}$$

The answer is (B).

Author Commentary

🕐 Use EPA-recommended exposure factors. Tabulating the exposure factors used will save time.

💣 Risk is expressed as a likelihood (e.g., one in one million). Therefore, the units in the calculations must cancel. Lifetime risk is the risk from exposure to both chemicals and is additive.

71. Find the temperature of the mixed flows.

T_m temperature of the mixed flows   °C

$$T_m = \frac{\left(280 \; \frac{m^3}{s}\right)(6°C) + \left(11 \; \frac{m^3}{s}\right)(20°C)}{280 \; \frac{m^3}{s} + 11 \; \frac{m^3}{s}}$$
$$= 6.5°C$$

For temperatures between 4°C and 20°C, the temperature variation constant is 1.135.

k deoxygenation rate constant d^{-1}
θ temperature variation constant   –

$$k_{6.5} = k_6 \theta^{6.5°C - 6°C} = \left(\frac{0.080}{d}\right)(1.135)^{6.5°C - 6°C}$$
$$= 0.085 \; d^{-1}$$

The answer is (B).

Author Commentary

🕐 Solve the problem in two parts: first find the mixed flow temperature, second, apply the correction to the river constant. A temperature variation constant consistent with the temperature range given is required for the second part. A fairly wide range of values can be used without significantly altering the solution, so don't waste time trying to select the "best" value.

💣 It is easy to accidentally reverse the temperatures used in the exponent of the temperature variation constant. Review the answer. If the mixed flow temperature is greater than 6°C, the river constant should increase. If it does not increase, then the temperatures were likely reversed in the calculation.

72. Biochemical oxygen demand is the amount of dissolved oxygen (DO) needed to biologically oxidize organic matter. ThOD is the stoichiometric oxygen required to completely oxidize organic matter. Therefore, as the BOD to ThOD ratio approaches one for a particular waste, it becomes more likely that waste will be biologically degradable.

The answer is (C).

Author Commentary

🕐 The correct answer is the chemical with a ratio of BOD to ThOD that is closest to one, so the likely options are A or C. A quick calculation of these two ratios will reveal the correct option.

💣 The answer options suggest either calculating based on the magnitude of the BOD concentrations or the ratios of BOD to ThOD. The correct answer will be based on the ratios.

73. Because the lake has a constant volume, the flow rates in and out of the lake are assumed to be equal at 735,000 ft³/yr.

$NH_{3(in)}$ ammonia mass in inflow   mg/yr
$NH_{3(out)}$ ammonia mass in outflow   mg/yr
ΔNH_3 ammonia mass lost to nitrification   mg/yr

$$NH_{3(in)} = NH_{3(out)} + \Delta NH_3$$
$$\Delta NH_3 = NH_{3(in)} - NH_{3(out)}$$
$$= \left(1.2 \; \frac{mg}{L} - 0.26 \; \frac{mg}{L}\right)$$
$$\times \left(735{,}000 \; \frac{ft^3}{yr}\right)\left(28.3 \; \frac{L}{ft^3}\right)$$
$$= 2.0 \times 10^7 \; mg/yr$$

Assume that the ammonia concentration is given in the customary form as N (NH_3 as N). The nitrogenous oxygen demand is 4.57 g O_2/g N.

NOD	nitrogenous oxygen demand for nitrification	g O_2/g N
OD_e	oxygen demand exerted	kg O_2/yr

$$OD_e = (\Delta NH_3)(NOD)$$
$$= \frac{\left(2.0 \times 10^7 \; \frac{mg \; NH_3 \; as \; N}{yr}\right)\left(4.57 \; \frac{g \; O_2}{g \; N}\right)}{\left(1000 \; \frac{mg}{g}\right)\left(1000 \; \frac{g}{kg}\right)}$$
$$= 91 \; kg \; O_2/yr$$

The answer is (C).

Author Commentary

🕐 This is a mass balance problem. It saves time to define the inputs and outputs to the lake in a simple mathematical format: input = output + change. Rearrange the equation to isolate the unknown term, then do the calculations for each term in sequence.

💣 The difficulty in this problem is applying the nitrogenous oxygen demand (NOD) for nitrogen, which converts nitrogen loading to oxygen loading. Find the value for NOD from references.

74. The minimum required tank capacity is determined using the slope of the average flow. To find the slope of the average flow, first define the average flow line. Pass a line through the ordinate to the cumulative volume at 24 h. Next, draw lines parallel to the average flow line that are also tangential to both maximum deviations from the average flow line. The vertical separation of the parallel lines represents the minimum required storage volume.

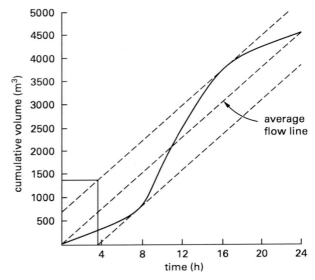

From the illustration, the vertical separation of the parallel tangent lines is

V	storage volume	m^3

$$V = 13\,000 \; m^3$$

The answer is (C).

Author Commentary

💣 Inflow and outflow occur simultaneously, so storage volumes may be smaller than intuitively expected. Storage volume is not the cumulative volume at the end of the 24 h period.

75. The initial and final concentrations are not given, but for 90% degradation, the ratio of initial concentration to final concentration is $C_o/C_i = 0.1$ by definition.

C_i	initial concentration	mg/L
C_o	final concentration	mg/L
k	reaction rate	h^{-1}
n	number of reactors in series	–
θ	residence time	h

$$\frac{C_o}{C_i} = \left(\frac{1}{1 + k\theta}\right)^n$$

$$0.1 = \left(\frac{1}{1 + \left(\frac{0.37}{h}\right)\theta}\right)^3$$

$$\theta = 3.17 \text{ h} \quad (3.2 \text{ h})$$

The answer is (C).

Author Commentary

🕐 The reaction rate units identify this as a first-order reaction problem. 90% removal is the same as 10% remaining, which means the ratio for the final to initial concentration is 0.1. Recognizing this relationship will save time.

💣 Because there are three reactors in series, the usual first-order equation is modified. The result is not the same as simply adding reaction times for each reactor.

76. From the graph, breakpoint chlorination occurs at a dose of 6.6 mg/L and a chlorine residual of 0.2 mg/L. A free chlorine residual of 0.2 mg/L occurs at an additional 0.2 mg/L above breakpoint. The corresponding chlorine dose is 9.2 mg/L.

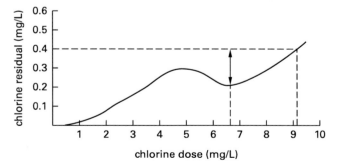

The answer is (D).

Author Commentary

🕐 There is no calculation required. Simply use the graph to find the dose corresponding to the desired residual.

💣 Do not confuse the free chlorine residual with the combined chlorine residual.

77. Convert Ca^{++} and HCO_3^- from mass concentrations to molar concentrations. The molecular weight of calcium is 40 g/mol.

C	concentration	mg/L
C_{molar}	molar concentration	mol/L
MW	molecular weight	mg/mol

$$C_{molar,Ca^{++}} = \frac{C_{Ca^{++}}}{MW_{Ca^{++}}} = \frac{46 \frac{mg}{L}}{\left(40 \frac{g}{mol}\right)\left(1000 \frac{mg}{g}\right)}$$

$$= 0.00115 \text{ mol/L}$$

The molecular weight of HCO_3^- is

$$MW_{HCO_3^-} = MW_H + MW_C + 3MW_O$$

$$= 1 \frac{g}{mol} + 12 \frac{g}{mol} + (3)\left(16 \frac{g}{mol}\right)$$

$$= 61 \text{ g/mol}$$

$$C_{molar,HCO_3^-} = \frac{C_{HCO_3^-}}{MW_{HCO_3^-}} = \frac{133 \frac{mg}{L}}{\left(61 \frac{g}{mol}\right)\left(1000 \frac{mg}{g}\right)}$$

$$= 0.00218 \text{ mol/L}$$

$$pCa = pCa^{++} = -\log C_{molar,Ca^{++}}$$

$$= -\log\left(0.00115 \frac{mol}{L}\right)$$

$$= 2.94$$

$$pM = pHCO_3^- = -\log C_{molar,HCO_3^-}$$

$$= -\log\left(0.00218 \frac{mol}{L}\right)$$

$$= 2.66$$

LSI — Langelier stability index

$pK'_2 - pK'_s$ — ionic strength and temperature constant

$$LSI = pH - \left((pK'_2 - pK'_s) + pCa + pM\right)$$
$$= 7.4 - (2.28 + 2.94 + 2.66)$$
$$= -0.48$$

The answer is (B).

Author Commentary

🕐 Throughout the solution, $px = -\log x$, where x is any ion in units of moles per liter. Save time by converting the concentrations of Ca^{++} and HCO_3^- to moles per liter before calculating pCa and $pHCO^3$.

78. Find the volume of the tank used for the flash mixer.

Q	flow rate	m^3/s
t_d	contact time	s
V_{tank}	tank volume	m^3

$$V_{tank} = t_d Q = \left(\frac{90 \text{ s}}{86\,400 \frac{s}{d}}\right)\left(20\,000 \frac{m^3}{d}\right)$$
$$= 20.83 \text{ m}^3$$

The absolute viscosity of water at 10°C is 1.3077×10^{-3} Pa·s.

G	minimum velocity gradient	s^{-1}
P	power requirement	kW
μ	absolute viscosity	Pa·s

$$P = \mu G^2 V_{tank}$$
$$= \frac{(1.3077 \times 10^{-3} \text{ Pa·s})(960 \text{ s}^{-1})^2(20.83 \text{ m}^3)}{1000 \frac{W}{kW}}$$
$$= 25.11 \text{ kW} \quad (25 \text{ kW})$$

The answer is (B).

Author Commentary

💣✳ This is a classic velocity gradient problem with a straightforward solution. The greatest potential for error is in converting to kW, so check these units carefully.

79. Convert the allowable remaining hardness into millimoles per liter. The molecular weight of calcium carbonate is 100 mg/mmol.

C	concentration	kg/d, mg/L, or mmol/L
MW	molecular weight	mg/mmol

$$C_{hardness,molar} = \frac{C_{CaCO_3}}{MW} = \frac{65 \frac{mg}{L}}{100 \frac{mg}{mmol}}$$
$$= 0.65 \text{ mmol/L}$$

$$C_{Ca^{++},removed} = C_{Ca^{++},initial} - C_{hardness,molar}$$
$$= 2.8 \frac{mmol}{L} - 0.65 \frac{mmol}{L}$$
$$= 2.15 \text{ mmol/L}$$

Sodium hydroxide reacts with calcium bicarbonate hardness as shown.

$$[Ca^{++} + 2HCO_3^-] + 2NaOH \rightarrow CaCO_{3(s)} + 2Na^+$$
$$+ CO_3^- + 2H_2O$$

2.15 mmol/L of Ca^{++} reacts with $(2)(2.15 \text{ mmol/L}) = 4.3$ mmol/L of HCO_3^- and $(2)(2.15 \text{ mmol/L}) = 4.3$ mmol/L of sodium hydroxide.

$$C_{HCO_3^-,remaining} = C_{HCO_3^-,initial} - C_{HCO_3^-,reacting}$$
$$= 5.7 \frac{mmol}{L} - 4.3 \frac{mmol}{L}$$
$$= 1.4 \text{ mmol/L}$$

Sodium hydroxide reacts with magnesium bicarbonate hardness as shown.

$$[Mg^{++} + 2HCO_3^-] + 4NaOH \rightarrow Mg(OH)_{2(s)} + 4Na^+$$
$$+ 2CO_3^- + 2H_2O$$

0.7 mmol/L of Mg^{++} reacts with $(2)(0.7 \text{ mmol/L}) =$ 1.4 mmol/L of bicarbonate and $(4)(0.7 \text{ mmol/L}) =$ 2.8 mmol/L of sodium hydroxide.

$$C_{HCO_3^-,\text{after } Mg^{++}} = C_{HCO_3^-,\text{remaining}} - C_{HCO_3^-,\text{reacting}}$$

$$= 1.4 \, \frac{\text{mmol}}{\text{L}} - 1.4 \, \frac{\text{mmol}}{\text{L}}$$

$$= 0 \text{ mmol/L}$$

$$C_{Mg^{++},\text{remaining}} = C_{Mg^{++},\text{initial}} - C_{Mg^{++},\text{reacting}}$$

$$= 1.9 \, \frac{\text{mmol}}{\text{L}} - 0.7 \, \frac{\text{mmol}}{\text{L}}$$

$$= 1.2 \text{ mmol/L}$$

Sodium hydroxide reacts directly with magnesium non-carbonate hardness as shown.

$$Mg^{++} + 2NaOH \rightarrow Mg(OH)_{2(s)} + 2Na^+$$

1.2 mmol/L of Mg^{++} reacts with $(2)(1.2 \text{ mmol/L}) =$ 2.4 mmol/L of sodium hydroxide.

The total sodium hydroxide required is

$$C_{NaOH,\text{total required}} = 4.3 \, \frac{\text{mmol}}{\text{L}} + 2.8 \, \frac{\text{mmol}}{\text{L}} + 2.4 \, \frac{\text{mmol}}{\text{L}}$$

$$= 9.5 \text{ mmol/L}$$

$$MW_{NaOH} = 23 \, \frac{\text{mg}}{\text{mmol}} + 16 \, \frac{\text{mg}}{\text{mmol}} + 1 \, \frac{\text{mg}}{\text{mmol}}$$

$$= 40 \text{ mg/mmol}$$

The total amount of sodium hydroxide required per day is

Q flow rate m^3/d

$$C_{NaOH,\text{total}} = C_{NaOH,\text{total required}}(MW_{NaOH}) Q$$

$$= \left(9.5 \, \frac{\text{mmol}}{\text{L}}\right)\left(40 \, \frac{\text{mg}}{\text{mmol}}\right)\left(12\,000 \, \frac{\text{m}^3}{\text{d}}\right)$$

$$\times \left(10^{-6} \, \frac{\text{kg}}{\text{mg}}\right)\left(1000 \, \frac{\text{L}}{\text{m}^3}\right)$$

$$= 4560 \text{ kg/d} \quad (4600 \text{ kg/d})$$

The answer is (B).

Author Commentary

🕐 Find the reaction equations for sodium hydroxide softening, and work through each reaction sequentially, calculating the sodium hydroxide required for each one.

💣 Using concentration units of millimoles per liter will allow easy addition, subtraction, and multiplication of chemical quantities. Using any other units will unnecessarily complicate the solution or, in the case of units of milligrams per liter, will give an incorrect answer.

80. Find the weighted rating for each site.

R	rating	—
WF	weighting factor	—
WR	weighted rating	—

$$WR = (WF)R$$

criteria	WF	site 1 R	site 1 WR	site 2 R	site 2 WR
haul distance	2	2	4	3	6
access routes	4	3	12	1	4
land value	3	2	6	2	6
permeability	3	3	9	4	12
heterogeneities	4	2	8	3	12
cover quantities	3	3	9	2	6
seismic activity	4	4	16	4	16
quality	2	3	6	3	6
gradient	1	2	2	3	3
depth	2	4	8	3	6
drainage pattern	3	3	9	2	6
streams	2	3	6	2	4
population	3	2	6	3	9
land uses	4	4	16	3	12
opposition	4	3	12	2	8
	44		129		116

WA   weighted average

$$WA = \frac{\sum WR}{\sum WF}$$

$$WA \text{ site } 1 = \frac{129}{44} = 2.9$$

$$WA \text{ site } 2 = \frac{116}{44} = 2.6$$

$$2.9 > 2.6$$

Choose site 1, which has a weighted average of 2.9.

The answer is (C).

Author Commentary

🕐   This problem is solved using a weighted average analysis. The most direct way of applying the analysis is to add columns to the table for the weighted rating (WR) for each site. This will facilitate quickly multiplying the WF and R columns (a calculation simple enough to be done without a calculator) and adding the columns.

💣   The option with the greatest, not the smallest, weighted average score is the best option.